ORGANIC GEOCHEMISTRY FOR EXPLORATION GEOLOGISTS

Douglas Waples
Colorado School of Mines

**Burgess Publishing Company
CEPCO Division**

Cover: A scanning electron micrograph of calcareous nannoplankton from a Gulf Coast sediment, Upper Miocene age (magnification 10,000 times). Top to bottom: *coccolithus pelagicus, discoaster brouweri, d. variabilis,* and *d. challengeri.* These and other phytoplanktonic organisms provide much of the organic material preserved in sediments, and are therefore the source materials for petroleum.

Managing Editor: Judith Goodrich

Copyright © 1981 by Burgess Publishing Company
Printed in the United States of America
Library of Congress Catalog Number: 80-70001
ISBN: 0-8087-2961-6 (cloth)
      0-8087-2980-2 (paper)

Burgess Publishing Company
7108 Ohms Lane
Minneapolis, Minnesota 55435

All rights reserved. No part of this book may be reproduced in any form whatsoever, by photograph or mimeograph or by any other means, by broadcast or transmission, by translation into any kind of language, nor by recording electronically or otherwise, without permission in writing from the publisher, except by a reviewer, who may quote brief passages in critical reviews and articles.

H G F E D C B A

BOARD OF CONSULTING EDITORS FOR GEOSCIENCE

Board of Consulting Editors for Geoscience

Series Editor: George deVries Klein
Department of Geology
University of Illinois at Urbana-Champaign
Urbana, IL 61801

Consulting Editors:

George Brimhall
Department of Geology and Geophysics
University of California-Berkeley
Berkeley, CA 94720

William J. Hinze
Department of Geosciences
Purdue University
West Lafayette, IN 47907

W. Stuart McKerrow
Department of Geology and Mineralogy
University of Oxford
Parks Road
Oxford OX1 3PR United Kingdom

J. Casey Moore
Earth Sciences
University of California-Santa Cruz
Santa Cruz, CA 95064

Peter A. Scholle
Branch of Oil and Gas Resources
United States Geological Survey
Denver, CO 80225

To Susan, who helped me learn.

Acknowledgments

Many people have encouraged my interest in and enthusiasm for petroleum geochemistry. Keith Kvenvolden first showed me how organic chemistry and geology could be combined. Dietrich Welte gave me my earliest opportunity to carry out research in organic geochemistry, and instilled in me a recognition of the importance of always maintaining a proper geological perspective in any geochemical problem. I thank the Alexander von Humboldt Foundation for making my tenure with Professor Welte's group possible.

Latin American Teaching Fellowships, a private foundation, supported me for a year in Valparaíso, Chile, and thus enabled me to gain my first experience in source rock analysis with the Empresa Nacional de Petróleo (ENAP). This period coincided with the last year of Salvador Allende's presidency, and provided a fascinating introduction to Latin America.

I especially thank Bob Jones and many other geologists with Chevron U.S.A. for the intensive education I received from them. It is, after all, the exploration geologist who finally decides whether organic geochemistry is used effectively in petroleum exploration.

Finally, I thank my students and colleagues at the Colorado School of Mines, especially, Claudia True, for the opportunity to test my teaching ideas on them.

Douglas Waples
Golden, Colorado
May 10, 1980

CONTENTS

Introduction 1

1. Brief Review of Organic Chemistry, 4

2. The Carbon Cycle and Preservation of Organic Material, 11

3. Composition of Kerogens, Bitumens, Petroleums, and Natural Gases, 20
   1. Introduction, 20
   2. Kerogen, 21
   3. Petroleum and Bitumen, 23
   4. Solidified Bitumens, 30
   5. Natural Gas, 30

4. Catagensis of Organic Material and the Formation of Oil and Gas, 32
   1. Introduction, 32
   2. Kinetics of Catagenesis, 32
   3. Bitumen Composition, 36
   4. Gaseous and Gasoline-Range Hydrocarbons, 42
   5. Kerogen Structure and Composition, 42

5. Migration of Oil and Gas, 47
   1. Introduction, 47
   2. Mechanisms of Primary Migration, 48
   3. Timing of Primary Migration, 53
   4. Mechanisms of Secondary Migration, 54
   5. Accumulation, 55
   6. Accumulation Efficiency, 57

6. Sample Analysis, 58
   1. Introduction, 58
   2. Sample Quality, 58
   3. The Analytical Scheme, 59

7. Data Interpretation, 66
1. Introduction, 66
2. Aspects of the Oil-Source Potential Problem, 66
3. Direct Methods for Measuring Oil-Generative Capacity, 68
4. Indirect Methods for Estimating Oil-Generative Capacity, 69
5. Evaluation of Oil-Source Capacity, 75
6. Application to Source Rock Analysis, 77
7. Evaluation of Gas Source Capacity, 83
8. Determining Growth and Depositional Environments, 86
9. Correlations, 88

8. Time and Temperature as Factors in Oil Generation, 95
1. Introduction, 95
2. Construction of the Geologic Model, 96
3. Special Cases, 99
4. Theory of Lopatin's Method, 100
5. Calculation of TTI Values, 101
6. Interpretation of TTI Values, 102
7. Correlation of TTI with other Geochemical Data, 103
8. Application of TTI Data to Exploration, 104

9. Practice Problems, 107
1. Source-Rock Evaluation, 107
2. Oil-Oil Correlations, 119
3. Oil-Source Rock Correlations, 127
4. Lopatin Reconstructions, 131

References Cited, 138

Subject Index, 143

INTRODUCTION

The application of geochemical techniques to oil and gas exploration has only recently achieved widespread acceptance among exploration geologists. In the past five years, efforts of the last half century in many diverse fields have been combined to give a reasonably clear picture of the chemical processes involved in petroleum formation. Organic geochemists are now able to make reasonably good semiquantitative predictions of the probability of sedimentary rocks containing oil.

Proponents of the organic origin of petroleum have shown in general how plant debris is converted to oil, but the transformation process is very complex, and not all the details have been fully elucidated. It is known that dead plant material, mainly from microscopic organisms called algae, is deposited in fine-grained sediments. During transport to the site of deposition, and for a short time (geologically speaking) thereafter, chemical and microbial transformations destroy some of the organic matter and alter the chemical composition of the remainder. Organic geochemists call these low-temperature transformations *diagenesis*.

The organic matter that emerges as a result of diagenetic reactions consists of many types of molecules whose sizes range from very small to immensely large. The large molecules are called *kerogen*; they play a key role in petroleum formation.

As the depth at which this organic material is buried increases, porosity and permeability decrease and temperature increases. These changes lead to a gradual cessation of microbial activity, bringing diagenesis to a halt. As temperature rises, however, thermal reactions become increasingly important. During this phase, which is called *catagenesis*, kerogen begins to decompose into smaller, more mobile molecules which can migrate out of the fine-grained source rock and into more porous and permeable conduits. These conduits often lead to traps, where migration ceases and oil accumulation occurs.

The goal of an organic geochemist who is engaged in pure research is to understand the molecular basis of diagenetic and catagenetic processes. An explorationist, however, has very different objectives in mind: he wants a tool that will help him to reduce the risk involved in searching for petroleum. In essence, he needs a quantitative measure of the oil-generating capacity of sedimentary rocks on a regional scale. He need not understand all the details of bacterial metabolism, but he should have a general appreciation of how bacterial effects will differ from one depositional environment to another, and of the implications of these differences vis-à-vis oil-generating capacities. He need not concern himself with the specific reactions involved in the catagenetic breakdown of kerogen, but he must understand the quantitative relationship between catagenesis and oil generation. He need not understand the physical chemistry of migration processes, but he should be able to evaluate the relative migrational efficiencies of different oil provinces.

Developing the skills necessary to carry out intelligent and useful source rock evaluations is not difficult, even if one does not have a strong chemistry background. This text is intended to foster the development of such skills. In this context, it is perhaps appropriate to define a few terms and to give some historical perspective to petroleum geochemistry.

It has been noted that the organic molecules in a sediment or a sedimentary rock come in a wide range of molecular sizes. The smaller molecules are usually soluble in common organic solvents. If a crushed rock sample is extracted with an organic solvent, the *extractable organic material (EOM)*, also called *bitumen,* can be recovered from the solvent. The insoluble organic material is called kerogen. Kerogen molecules are large; they are random polymers formed from many kinds of smaller reactive organic molecules present in fine-grained sediments. EOM usually represents 5-10% of the total organic matter in fine-grained sedimentary rocks; the majority is kerogen. The small molecules produced by catagenesis belong to the EOM fraction.

It is the bitumen molecules, which are small and therefore relatively mobile, that eventually migrate and sometimes accumulate in reservoirs as petroleum. Bitumen is therefore similar to petroleum chemically, and can be analyzed by the same techniques.

Detailed analyses of bitumen and petroleum require sophisticated separation techniques, because both are complex mixtures of many distinct chemical compounds. Before about 1960 these separations were not possible on a routine basis, and bitumen and petroleum analyses were therefore not employed in exploration programs.

By the early 1960's, however, the gas chromatograph had been almost universally adopted by petroleum geochemists. These relatively simple and inexpensive instruments facilitated analysis of bitumen and petroleum on the molecular level, and enabled workers to isolate and identify many new trace components of crude oils. It became possible, therefore, to correlate crude oils with each other and with bitumens.

Unfortunately, the compounds which are most easily studied by gas chromatography are also the simplest compounds chemically. As such they convey less information about the origin and history of crude oils than do the more complex (but more difficult to analyze) molecules.

As organic geochemists sought to expand their horizons, it was natural that their attention focused on kerogen. It is, after all, the dominant organic component of sedimentary rocks. But more importantly, it had become apparent that kerogen is the precursor of petroleum.

Organic geochemists began to wonder whether the properties of the kerogen found in sedimentary rocks could be used as a measure of the degree to which petroleum formation had occurred in those rocks and could occur there in the future. It became apparent that there already existed a wealth of data and several well-developed methodologies that could be applied to kerogen analyses. Gas chromatography could not be used for analyzing kerogens because of their high molecular weights, but numerous other approaches had already been worked out by coal scientists.

Since the early 1900's, chemists had studied the chemical, physical, and optical properties of coals with the goal of understanding and predicting their behavior as fuels. When it was realized that coals and kerogens had a great deal in common, the coal chemists' methods and terminology were rapidly appropriated by petroleum geochemists. Transmitted and reflected light microscopy, elemental analysis, and pyrolysis, all common methods for analyzing kerogens today, have their roots in pioneering work done by coal scientists. As a result, techniques for analyzing kerogen have developed rapidly in the last decade, and are of paramount importance today in understanding the oil-generative history of sedimentary rocks.

A coherent picture has now emerged of the three factors—kerogen quantity, quality, and thermal maturity—which determine the oil source capability of a sedimentary rock. We now are in a position to make at least semiquantitative predictions about the amounts of petroleum which have been or will be generated by a given volume of source rock. This is a most remarkable advance.

Unfortunately, knowledge of the amount of petroleum generated from a source rock solves only one part of the problem of finding petroleum reservoirs. Migration from source rock to reservoir must also occur. Organic geochemical studies of source rocks do not tell us much about migrational efficiency, and our poor understanding of the mechanisms of oil migration is another severe handicap. Much more research will be necessary before we are able to treat migrational processes quantitatively.

Although we cannot evaluate migrational efficiency even semiquantitatively, we still know that oil must be generated before it can migrate. The oil-generating capacity of a proposed source rock is an upper limit to the amount of oil which can finally be emplaced in a reservoir, and source-rock analyses can therefore be used in a regional exploration program to identify relatively high- and low-risk exploration areas.

In the last decade, significant progress has been made toward resolving the age-old dilemma of how time and temperature can be substituted for each other in the process of petroleum formation. Such questions as, "Can oil be generated from deep, hot Pleistocene sediments?" can now be answered with some certainty, thanks to pioneering work by Lopatin in the Soviet Union.

What will be the major developments in petroleum geochemistry in the 1980's? My guesses are the following:

1. Continued refinement of kerogen analysis as it is applied to evaluating oil-generating capability of sedimentary rocks. Quantitative evaluation will become increasingly popular.
2. Resurgence of interest in bitumen analysis for oil–oil and oil–source rock correlations. The availability of highly sophisticated combined gas chromatograph–mass spectrometer–computer systems will allow detailed analysis of micro- and nanogram quantities of complex molecules that can be used as sensitive tracers. Applications of these techniques to exploration are still rare, but the results of recent research appear promising.
3. Development of quantitative models of the migrational processes by which bitumen moves from the source rock to the reservoir. This development has been needed for a long time, and may finally be possible as a result of recent improvements in our understanding of oil generation and migration.
4. Development of predictive models for oil generation on a regional scale, based on time and temperature considerations. This advance could prove to be the most important step forward in the history of organic geochemistry.

One final word of clarification is in order here. The field that we call organic geochemistry is highly interdisciplinary, and many of the analytical techniques employed by petroleum geochemists are not strictly chemical. (Good examples of such nonchemical approaches are transmitted- and reflected-light microscopy.) Because these nonchemical methods provide valuable information about the nature of the organic material in sedimentary rocks, they are conveniently lumped with standard chemical procedures as part of the geochemist's repertoire.

1 | BRIEF REVIEW OF ORGANIC CHEMISTRY

Anyone whose work involves petroleum geochemistry must be able to communicate to some degree in the chemical idiom. It is therefore necessary that he or she become acquainted with some of the nomenclature of organic chemistry. Only a few of the myriad types of organic compounds are important components of petroleum, and we discuss those classes of compounds here.

In chemical terms, a *hydrocarbon* is a compound that contains only the elements carbon and hydrogen. Petroleum itself sometimes is referred to (not by chemists!) as "hydrocarbons," as in a geologist friend's favorite description of oil exploration as, "the search for the elusive hydrocarbon." Although petroleum does contain a large proportion of hydrocarbons, it often also contains substantial amounts of nitrogen, sulfur, oxygen, and other elements.

Examples of simple hydrocarbons are methane, ethane, and cyclohexane; their structures are shown below.

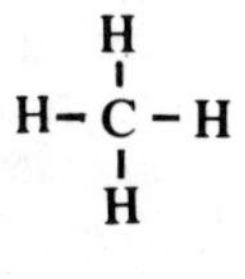

Methane 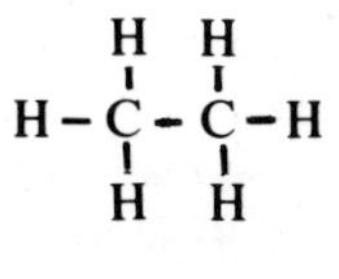   Ethane 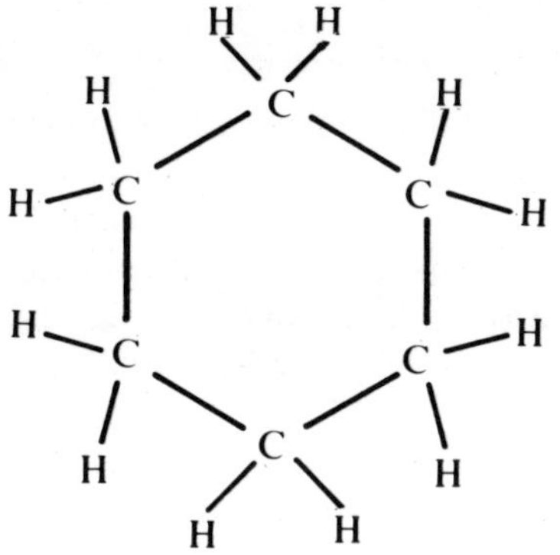   Cyclohexane

Note that in each of these compounds every carbon atom forms four bonds. This is always true in any organic compound, and we may thus use several different shorthand notations to relieve the drudgery of writing down all the hydrogen atoms. One common convention is to represent all the hydrogens attached to a given atom by a single H, using a subscript to denote the total number of hydrogens around that atom. The molecular structures of methane, ethane, and cyclohexane are thus represented by

CH₄
Methane

CH₃CH₃
or
H₃CCH₃
Ethane

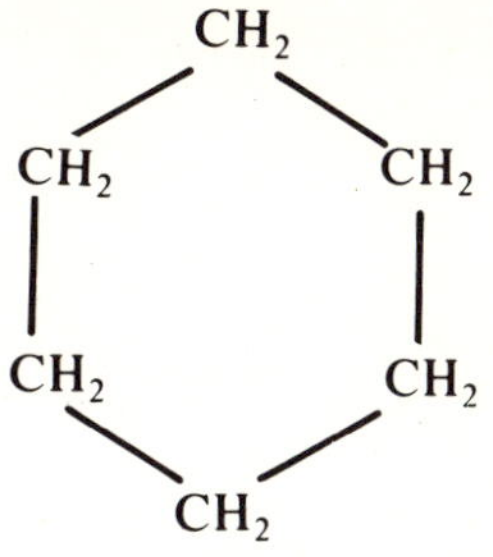

Cyclohexane (hexagonal ring of CH₂ groups)

We can make other logical simplifications for longer carbon chains. Thus the following representations of the molecule pentane are equivalent.

$$CH_3CH_2CH_2CH_2CH_3 \quad \text{or} \quad CH_3(CH_2)_3CH_3$$

Pentane

An even quicker shorthand that uses no letters has evolved. Each carbon atom is represented by a point, and carbon–carbon bonds are shown as lines connecting these points. Because we know that each carbon atom always forms four bonds, simple inspection indicates how many hydrogens each carbon atom must have. Thus pentane and cyclohexane are represented by the simplified structures shown below.

Pentane    Cyclohexane

The zig-zag configuration (illustrated by pentane) is adopted to show clearly each carbon atom.

The simplest series of hydrocarbons has linear structures; such molecules are called *n*-paraffins or *n*-alkanes. The letter *n* stands for "normal," and means that there is no branching in the carbon chain. The names of the ten simplest *n*-alkanes and three ways of representing each of them are given in Table 1.1. Note that the ending of each name is -ane,* as in "alkane." The first four names are irregular, but the prefixes denoting the number of carbon atoms in the other alkanes are derived from the Greek.

Carbon atoms need not always bond together in a linear arrangement. *Branching* can occur, giving a vast number of possible structures. Two of the many compounds of formula C_7H_{16} are shown below.

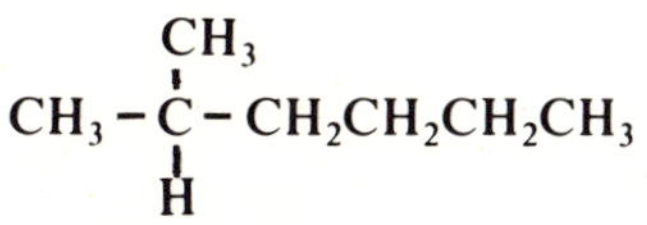

$$CH_3-\underset{\underset{H}{|}}{\overset{\overset{CH_3}{|}}{C}}-CH_2CH_2CH_2CH_3 \qquad CH_3-\underset{\underset{H}{|}}{\overset{\overset{H_3C}{|}}{C}}-\underset{\underset{CH_3}{|}}{\overset{\overset{CH_3}{|}}{C}}-CH_3$$

2-Methylhexane 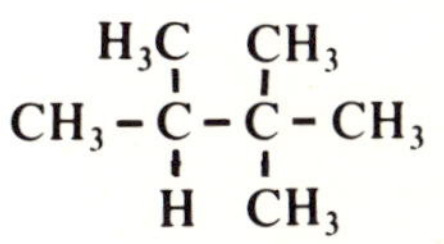   2,2,3-Trimethylbutane

---

*The suffix *ane* indicates that the hydrocarbon molecules are *saturated*, that is, they possess no multiple (double or triple) bonds (*vide infra*).

Table 1.1. Names and Abbreviations for n-Paraffins

Name	Abbreviations		
Methane	CH_4	CH_4	None
Ethane	C_2H_6	CH_3CH_3	None
Propane	C_3H_8	$CH_3CH_2CH_3$	∧
Butane	C_4H_{10}	$CH_3(CH_2)_2CH_3$	 ∧∨
Pentane	C_5H_{12}	$CH_3(CH_2)_3CH_3$	 ∧∨∧
Hexane	C_6H_{14}	$CH_3(CH_2)_4CH_3$	 ∧∨∧∨
Heptane	C_7H_{16}	$CH_3(CH_2)_5CH_3$	∧∨∧∨∧
Octane	C_8H_{18}	$CH_3(CH_2)_6CH_3$	 ∧∨∧∨∧∨
Nonane	C_9H_{20}	$CH_3(CH_2)_7CH_3$	 ∧∨∧∨∧∨∧
Decane	$C_{10}H_{22}$	$CH_3(CH_2)_8CH_3$	 ∧∨∧∨∧∨∧∨

Compounds that contain rings of carbon atoms can also be formed. One such compound, cyclohexane, has already been mentioned. These cyclic hydrocarbons are named by counting the number of carbon atoms in the ring and attaching the prefix *cyclo*. Examples are given below.

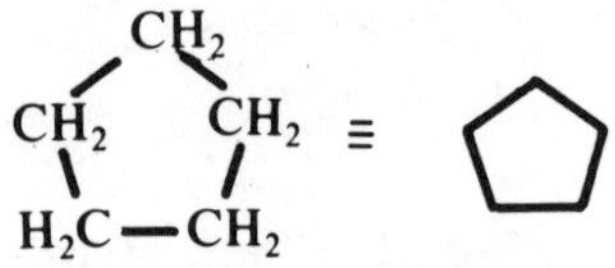

Cyclopentane 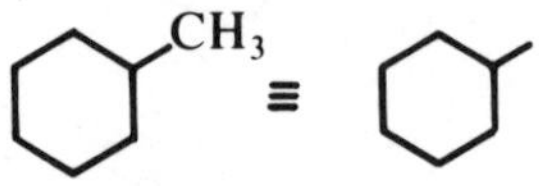   Methylcyclohexane

All of the compounds mentioned above are called *saturated hydrocarbons*, because they are saturated with respect to hydrogen. No more hydrogen can be incorporated into the molecule without breaking the molecule apart. Another important category of hydrocarbons is the *unsaturates*, which can combine with additional hydrogen. Many of these compounds have carbon-carbon double bonds; such compounds are called *alkenes*, or *olefins*. Examples are ethene, propene, and cyclohexene, whose structures are shown below. They are named in analogy to the corresponding alkanes.

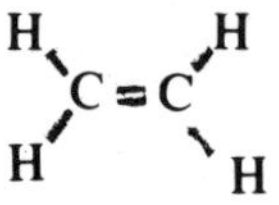

Ethene
(Ethylene)

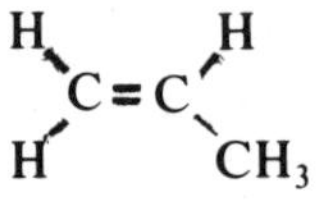

Propene
(Propylene)

Cyclohexene

Note that a carbon-carbon double bond is shown by two parallel lines instead of one. **Alkenes are converted to alkanes by the addition of hydrogen in the presence of a catalyst.** The hydrogenation of ethene to form ethane is shown below.

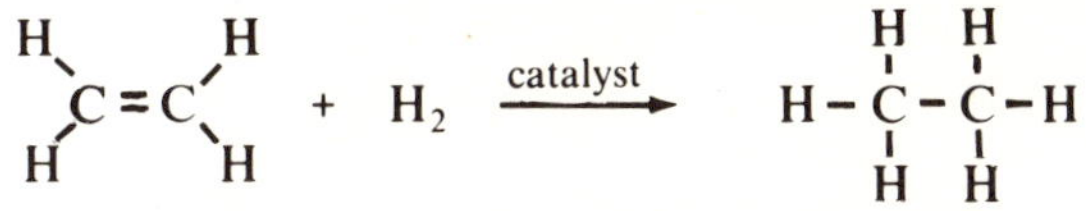

$$\underset{\text{Ethene}}{\overset{H}{\underset{H}{C}}=\overset{H}{\underset{H}{C}}} + H_2 \xrightarrow{\text{catalyst}} \underset{\text{Ethane}}{H-\overset{H}{\underset{H}{C}}-\overset{H}{\underset{H}{C}}-H}$$

Another extremely important class of unsaturated hydrocarbons is the *aromatics*. At first glance they appear to be just cyclic alkenes containing several double bonds, but actually they are unusually stable compounds and are chemically very different from alkenes. Some typical and important aromatics are shown and named below.

Benzene 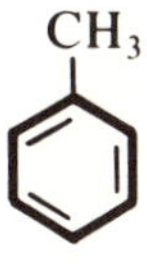   Toluene 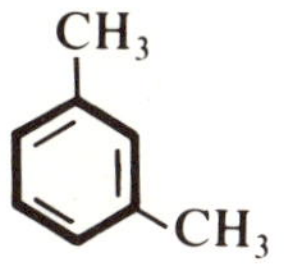   *m*-Xylene 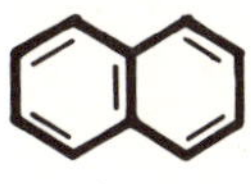   Naphthalene

Aromatics possess alternating single and double carbon-carbon bonds within a cyclic structure. A simplified notational form for these molecules consists of representing the double bonds by a circle within the ring. For example, the structures of toluene and naphthalene are represented by

Toluene 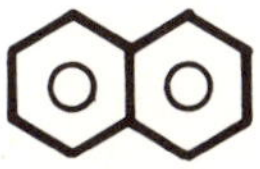   Naphthalene

The hydrocarbons that we have discussed are relatively small and simple molecules. They are important constituents of petroleum, but are very different from most of the organic molecules that are found in living organisms. Most molecules in living organisms are larger and more complex than the simple hydrocarbons, and they typically contain nitrogen, sulfur, phosphorus, and oxygen in functional groups (reactive parts of the molecule). The small hydrocarbons present in petroleum often are the end products of extensive degradation of biogenic molecules. In fact, some complex hydrocarbons that are found in fossil organic material can be related directly to individual biological precursors. Three such hydrocarbons, and the molecules from which they are derived, are shown in Figure 1.1.

Each of the three hydrocarbons shown in Figure 1.1 is representative of a geochemically important group of compounds. Phytane is an *isoprenoid* hydrocarbon; such compounds consist of a straight chain of carbon atoms with methyl (CH_3) groups attached to every fourth carbon atom. Isoprenoids containing from eight to forty carbon atoms have been found in petroleum and shales. Cholestane is representative of the *steranes,* which contain three six-membered rings and one five-membered ring linked in the manner shown in Figure 1.1. The third saturated hydrocarbon shown in Figure 1.1, 28,30-bisnorhopane, is a *triterpane*. All such compounds contain five six-membered rings linked together as shown in the figure. Many steranes and triterpanes have been found in petroleum and in sedimentary rocks.

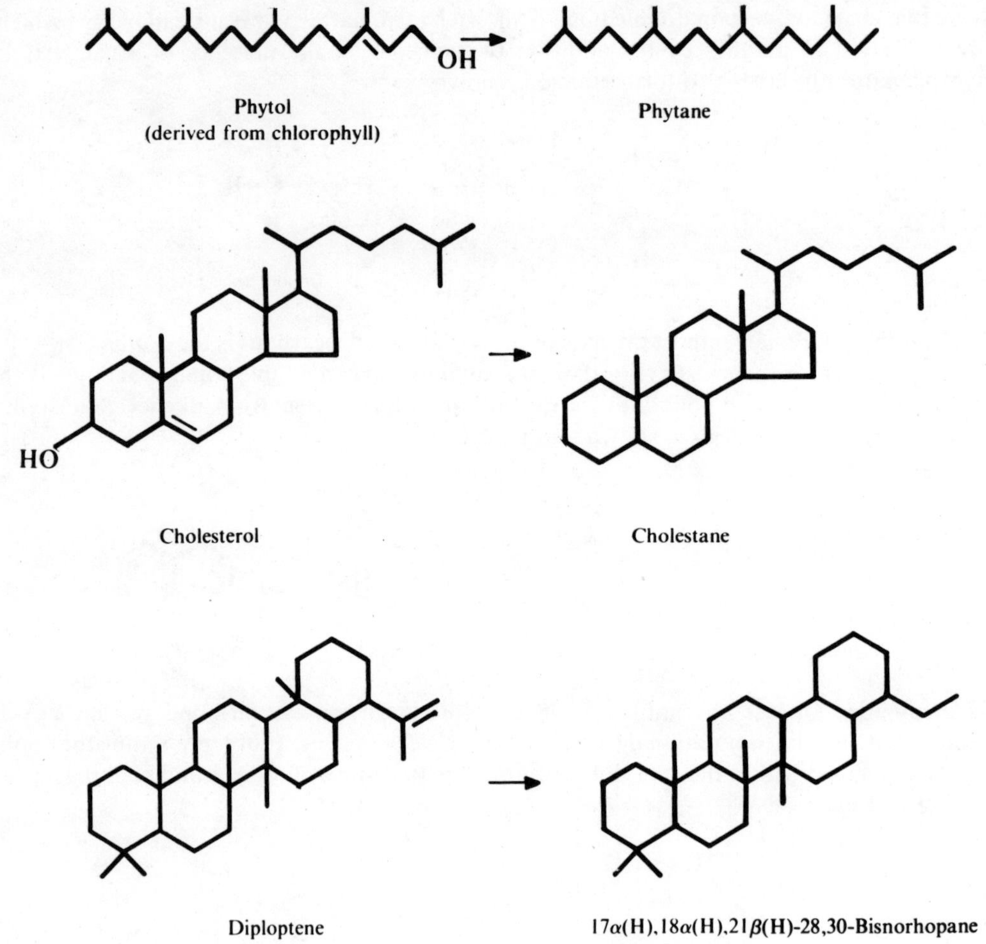

Phytol
(derived from chlorophyll)

Phytane

Cholesterol

Cholestane

Diploptene

17α(H),18α(H),21β(H)-28,30-Bisnorhopane

Figure 1.1. Representatives of some geochemically important groups of hydrocarbons and their biological precursors.

When they are part of an organic molecule, atoms other than carbon and hydrogen are referred to as *heteroatoms*. Molecules that contain heteroatoms are called *heterocompounds*. Nitrogen, oxygen, and sulfur are common constituents of organic molecules, and compounds that contain these heteroatoms are called *NSO compounds*. Fossil organic materials contain a large variety of heterocompounds, most of which are present in very small quantities.

The small amounts of heterocompounds present in fossil organic materials are difficult to isolate and analyze, but a few kinds are used interpretively in organic geochemistry. The first such heterocompounds to be studied extensively were the fatty acids. A molecule of a fatty acid consists of a long hydrocarbon chain that terminates in a carboxyl (COOH) group. A typical fatty acid is palmitic acid, whose structure is shown below.

$$CH_3(CH_2)_{14}-\overset{\overset{O}{\|}}{C}-OH$$

Palmitic acid
(hexadecanoic acid)

A second important class of heterocompounds is the *porphyrins*. These molecules are structurally related to, and in many cases derived from, chlorophyll. Chlorophyll is compared to a typical nickel porphyrin in Figure 1.2.

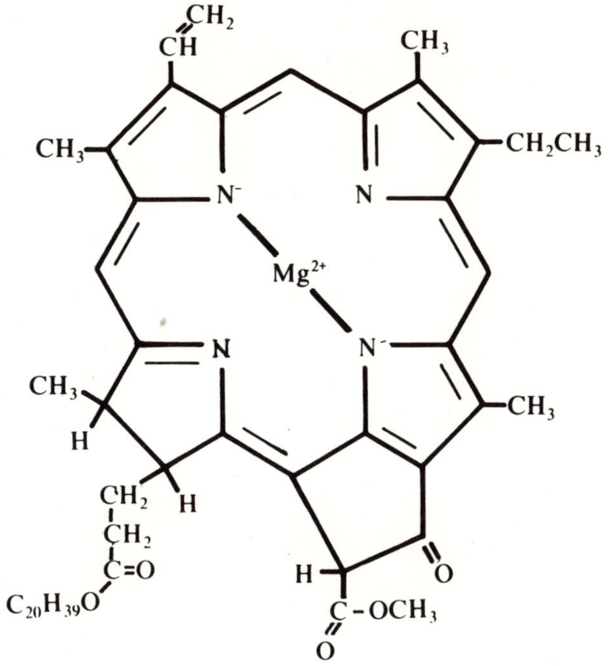

Chlorophyll a

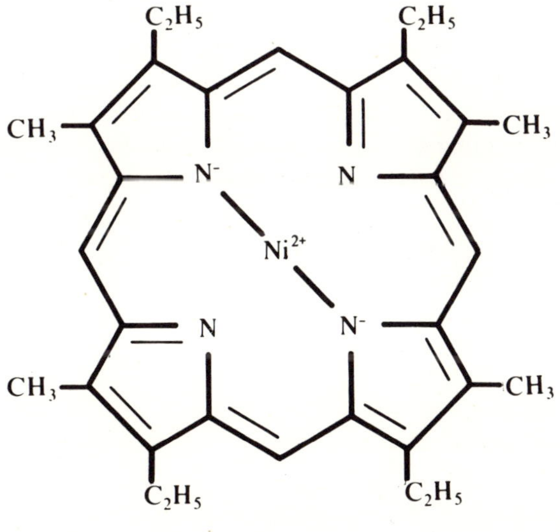

Nickel etioporphyrin II

Figure 1.2. Structures of chlorophyll and a typical nickel porphyrin.

There are many organic molecules that are not found in petroleum but are common components of other fossil organic material. Among these are lignin, carbohydrates, and amino acids.

Lignin is an important component of wood. It is a polymer that consists of many subunits (monomers) that have approximately the structure shown below.

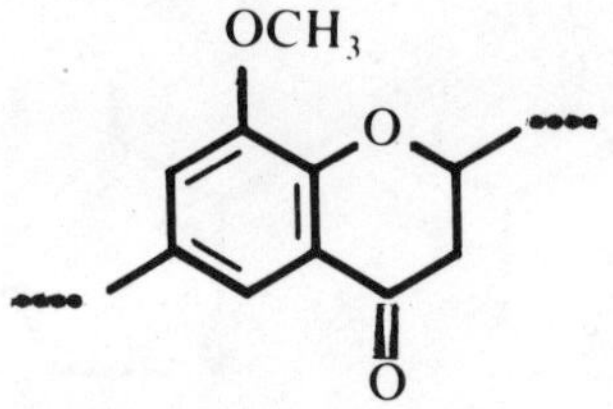

Lignin subunit (monomer)

The lignin monomers are linked together to form molecules that have molecular weights of from 3,000 to 10,000 atomic mass units (amu). When lignin decomposes it forms *phenols,* examples of which are given below.

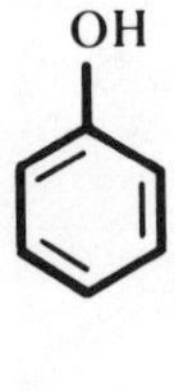

Phenol 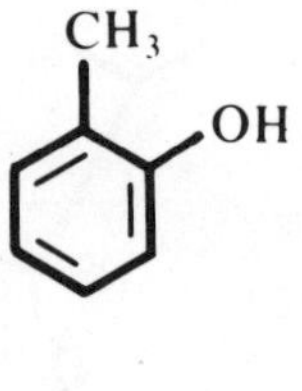   o-Cresol 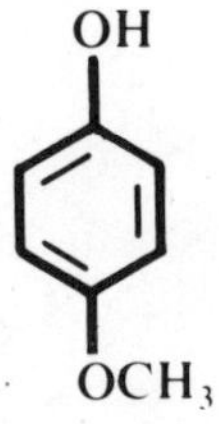   p-Hydroxyanisole

Phenols have a hydroxyl (OH) group attached to an aromatic ring. They are potent bacteriocides.

Cellulose is a polysaccharide, which means that it is a high-molecular-weight polymer composed of many simple sugar molecules. Cellulose has the general formula $(C_6H_{10}O_5)_n$, and belongs to the large class of compounds called *carbohydrates*. Cellulose is the most abundant organic compound in the biosphere.

Amino acids are the building blocks for proteins. Approximately twenty amino acids occur naturally in proteins, and each protein is constructed from a precise sequence of amino acids. Two common amino acids are serine and alanine, whose structures are shown below.

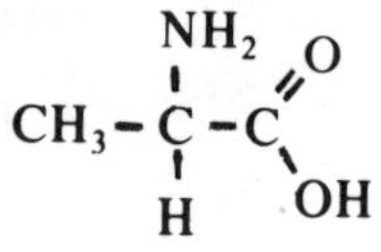

Alanine 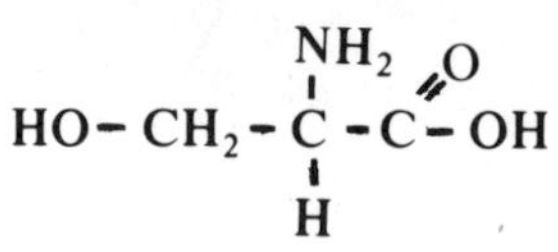   Serine

The listing in this chapter of classes of organic compounds is intended to provide a comfortable introduction to the terminology of petroleum geochemistry. Much more detailed discussions can be found in any introductory textbook of organic chemistry.

In the following chapters, we refer often to these classes of compounds and place them in a proper geochemical perspective.

2 | THE CARBON CYCLE AND PRESERVATION OF ORGANIC MATERIAL

Almost all life on earth is dependent upon the process of photosynthesis. Carbon dioxide (CO_2) is taken from the atmosphere by land plants, or from sea water by marine algae, and is then converted to plant tissues. Some plants die naturally; others are consumed by herbivores. The herbivores are in turn eaten by carnivores, and some of the carnivores are eaten by other carnivores. Even most dead organic material is consumed by scavengers. Photosynthesis is therefore the ultimate source of nourishment for nearly all living organisms.* It is clear, however, that this cannot be the whole story, for if it were our atmosphere would eventually become depleted of CO_2, and photosynthesis would cease. In order to avoid this depletion it is necessary for the carbon to be recycled by the conversion of plant and animal tissues back to atmospheric CO_2.

This recycling occurs in many ways. Animal and plant respiration returns CO_2 directly into the atmosphere. Bacterial decay and natural oxidation of dead animals and plants produce CO_2. The burning of fossil fuels by man and natural *in situ* combustion of coal seams and oxidation of oil seeps also recycle carbon.

The carbon cycle is not 100% efficient. Small amounts of organic material are continually dropping out of the cycle by being isolated in environments where oxidation to CO_2 cannot occur. It is probable that much less than 1% of the organic material produced escapes recycling (Garrels and Perry, 1974), but over geologic time this small flux has produced extremely large quantities of fossil organic material. Some of this fossil organic material is now stored as coal, oil, and natural gas. Thus the study of the origin of fossil fuels has as its primary concern that small fraction of organic carbon which escapes the carbon cycle.

Bolin (1970) has explained the carbon cycle very clearly in nontechnical terms. Figure 2.1, which represents the general aspects of the carbon cycle, is adapted from his article. It is immediately apparent that although less than a billion metric tons of organic carbon drops out of the carbon cycle each year, the process has been occurring for such a length of time that the total amount of preserved organic carbon has become incredibly large. Most of this carbon, however, is finely disseminated in sedimentary rocks and will never become concentrated enough to be economically recoverable. Not more than about 0.05% of the preserved organic material exists in commercial deposits, and most of this occurs as coal. Thus not more than one

*A few kinds of bacteria (for example, some bathypelagic species that live near undersea volcanic vents) are not dependent upon a photosynthetic driving mechanism for their metabolic processes. Such bacteria can be at the bottom of a nonphotosynthetically based food web. Nonphotosynthetic organisms were once the dominant forms of life on the earth; today they form the bottom of the food web only in very specialized ecological niches.

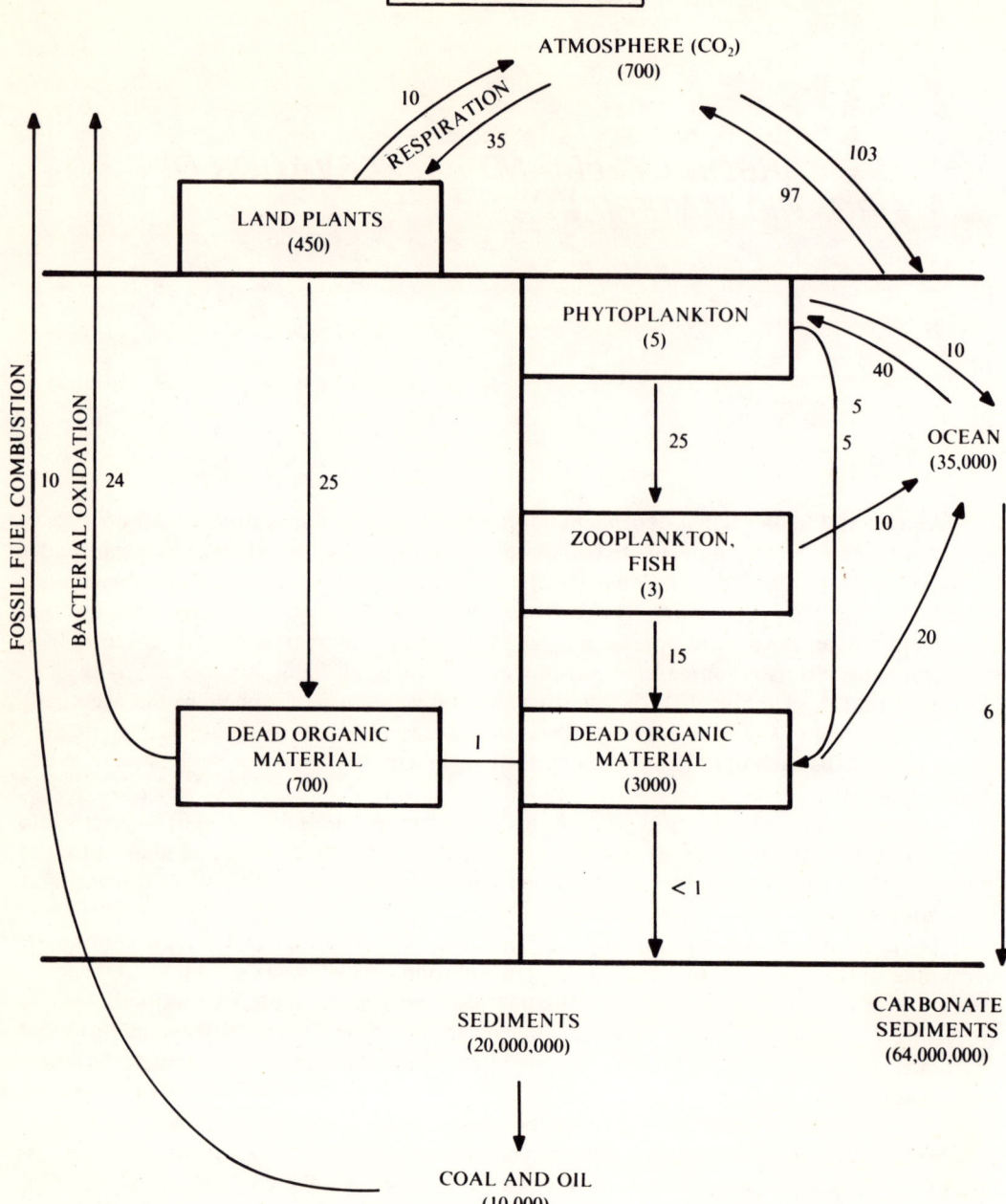

Figure 2.1. The carbon cycle. Numbers represent quantities in billions of metric tons. Numbers in parentheses represent stored quantities; numbers without parentheses are yearly fluxes. (Adapted from Bolin, *The Carbon Cycle*. Copyright © Scientific American, Inc. All rights reserved.)

CO_2 molecule out of every 1,000,000 taken up in photosynthesis eventually is converted to a fossil fuel. The series of events that is involved in the formation of an economically exploitable accumulation of petroleum is indeed highly selective.

Several other features of Figure 2.1 merit comment. Note that the total amount of carbon present at any one time in land plants (450 billion metric tons) is much greater than the amount stored in marine phytoplankton (microscopic photosynthetic plants such as algae; 5 billion metric tons), but the amount of CO_2 taken up annually by phytoplankton is actually greater than that taken up by land plants (40 billion versus 35 billion metric tons). This surprising situation results from the fact that, compared to land plants, phytoplankton have extremely short life cycles and high metabolic rates. These data also suggest that phytoplankton must have made a large contribution to the total organic carbon preserved in sediments.

But what are the details of the processes by which living organisms are transformed into coal, oil, or natural gas? This chapter is concerned only with low-temperature processes—those which occur at or near room temperature. The totality of such processes is called *diagenesis*. Thermal processes are discussed in Chapter 4.

When an organism dies, decay begins immediately. Complex molecules are broken down into smaller, simpler ones. A large part of most organisms consists of polymers of various kinds (e.g., cellulose, protein, chitin, sporopollenin), and so the process of decay may be described as a transformation of *biopolymers* into *geomonomers*. Examples of such transformations are shown in Figure 2.2.

Degradation of biopolymers can occur by both microbiological and nonbiological processes. Bacteria exist almost everywhere near the earth's surface, and all biopolymers are

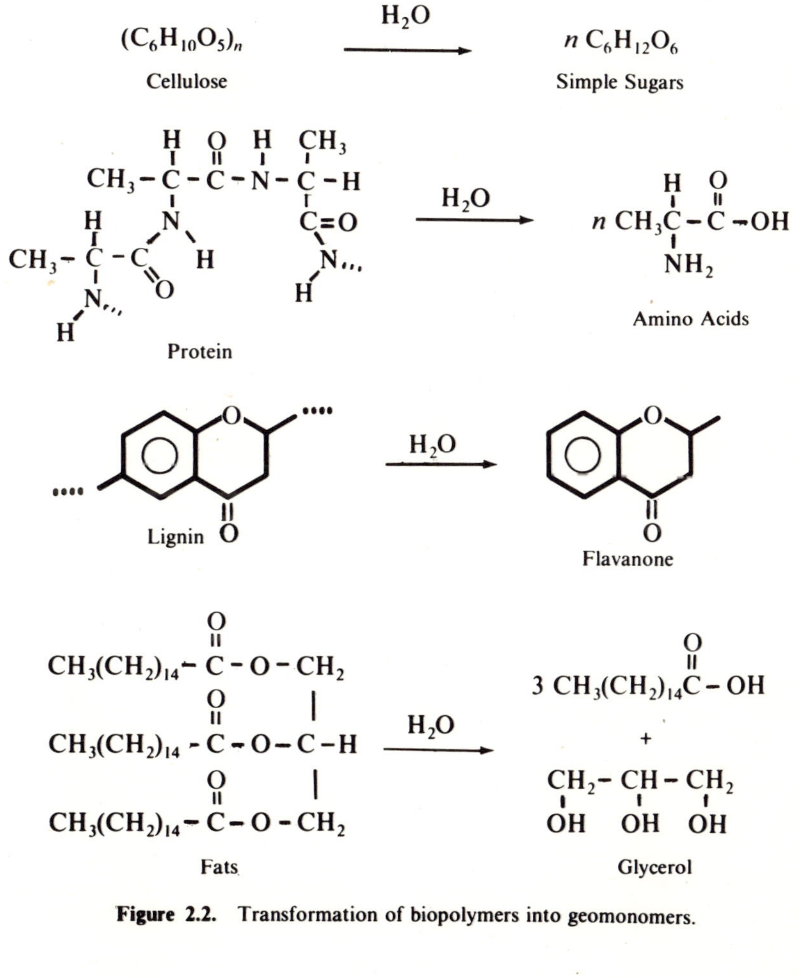

Figure 2.2. Transformation of biopolymers into geomonomers.

susceptible to attack by them. If bacterial activity proceded unchecked, there would be little or no preservation of organic material in sediments. But in fact there are many natural conditions that will hinder bacterial growth. Besides organic material, bacteria require trace metals and other nutrients; some require light; and most require a substrate that can accept the electrons released during oxidation of the organic material. For *aerobic bacteria,* oxygen is the electron acceptor; without molecular oxygen aerobes cannot metabolize nutrients. Other bacteria called *anaerobes* are killed or strongly inhibited in their actions by oxygen. Such bacteria require an oxygen-free environment such as that found in sapropelic muds.

Substances other than oxygen can be toxic to bacteria. The phenols, for example, are poisonous to many bacteria and are of geochemical interest because they are abundant in the waters in coal swamps. These waters sometimes actually are sterile enough to drink, even though they are colored brown by tannins from tree bark.

Because bacterial degradation seldom is able to proceed to completion, the breakdown from biopolymers to geomonomers is almost always unfinished to some degree. Furthermore, many of the geomonomers are themselves susceptible to rapid decomposition. For example, sugars and amino acids are consumed rapidly by microorganisms and so may be difficult to detect in sediments. Other compounds, such as phenols, hydrocarbons, and fatty acids, are more resistant to bacterial degradation.

At the same time that the biopolymers and geomonomers are being broken down, a process that is competitive with their degradation begins. Many of the molecules present in a dead organism are quite reactive chemically, and spontaneous reactions occur among them. These reactions result in an accumulation of molecules that have been synthesized from smaller ones. The products of the spontaneous reactions are constructed randomly from whatever organic debris happens to be present, and hence they are essentially random polymers. They are formed in the geosphere and so are called *geopolymers.* The random structures of geopolymers make them resistant to anaerobic bacterial degradation, a process which usually is highly specific. Geopolymers are, therefore, relatively stable, and their synthesis serves to preserve organic material even in the presence of bacteria.

There are several different kinds of geopolymers. Fulvic acids, humic acids, and kerogen are the most commonly studied, and of these, kerogen is of the greatest interest to petroleum geochemists.*

Kerogens derive from many different types of precursor molecules and are formed under a wide range of depositional conditions. There are therefore many different kinds of kerogens. Coal is best thought of as a special kind of kerogen that is relatively undiluted by mineral matter and that was formed from certain plant materials under specific geochemical conditions. Adoption of this approach helps to unify the concepts and processes of organic geochemistry.

The formation of kerogen occurs in two successive stages: polymerization and rearrangement. The polymerization stage involves the formation of geopolymers from geomonomers; it probably begins immediately upon an organism's death and is completed in a geologically short time—perhaps only a few hundreds or thousands of years. The rearrangement stage begins when the first geopolymers have been formed, and it continues for as long as the kerogen exists.

The degree to which organic material is preserved and kerogen is formed depends a great deal upon the biogeochemical conditions encountered by the organic material during transport, sedimentation, and burial. Table 2.1 lists several important types of depositional

---

*Fulvic and humic acids are derived chiefly from lignin, carbohydrates, and amino acids. They may themselves be converted to kerogen, but such humic kerogens are precursors of coaly material rather than petroleum.

environments and includes general descriptions of transport modes and depositional conditions. The last column of the table contains general predictions about the nature of any fossil fuels that might eventually be produced.

It is immediately apparent from Table 2.1 that terrestrial organic material will not normally generate significant amounts of oil. Low-energy environments where there is a minimal influx of mineral detritus are conducive to the development of excellent peat and coal swamps. In these environments nearly all the sediment may be organic. Coal swamps usually have extensive tree growth, and phenolic compounds in the wood act as bacteriocides that prevent bacterial decomposition of dead organic material. The large quantity of organic debris that is present consumes any oxygen that gets into the water, thus maintaining an anoxic environment. Conditions are ideal for the preservation of organic material. Stutzer (1932) has described in detail the geological and biological conditions present in coal swamps.

Coal swamps do not produce much oil, however, because the chemical composition of the organic material is wrong. Instead of being converted to hydrocarbons, the cellulose and lignin which make up the bulk of woody organic material are converted to peat, then to coal, and finally may yield some natural gas. Small amounts of fat and waxes, particularly those that occur in leaf coatings, pollens, and tree resins, may in some cases be converted to oil. Economically exploitable deposits of petroleum that are derived from these materials are rare, but they do exist. An area where coals and coal measures apparently have produced commercial petroleum accumulations is the Cooper Basin in Australia (Brooks et al., 1971).

Most terrestrial plants contain too little *lipid* material (fats, waxes, and oils) to form oil-generative kerogens. Cellulose and lignin are dominant, and from these we might expect that only small amounts of methane will be generated.

Most terrestrial material, however, does not end up in coal swamps. If it is not oxidized immediately by bacteria, it commonly is transported long distances to a final resting place in a coastal or deltaic depositional environment. This transport occurs in oxygenated water, and so results in much bacterial and nonbiological oxidation. Because oxidation lessens the possibility that the organic material will be converted into either oil or gas, most terrestrial organic material in coastal or deltaic environments is of very little interest to petroleum geochemists.

Any marine depositional environment will contain, in addition to allochthonous terrestrial debris, remains from the local flora, principally phytoplankton (algae). Algae do not contain lignin, and therefore do not leave resistant woody residues when degradation occurs. Where conditions are favorable for bacterial activity, algae are extensively degraded. Their lipid components are the most resistant to bacterial attack, and therefore will be preferentially preserved.

Bacteria use nutrients to promote their own growth and cell division. In other words, as organic material is degraded by microorganisms, some of it is converted into new tissues. Because many bacteria are rich in lipid components, there is a gradual conversion of plant carbohydrate and protein into bacterial lipids. This process leads to a gradual concentration of lipids in sediments during diagenesis. These lipids in turn can easily become the precursors for the hydrocarbon molecules found in petroleum.

The organic material in deltaic depositional environments is rather uniformly disseminated, with few occurrences of organic-rich layers. In many modern deltaic settings, quiet lagoons exist where organic material becomes concentrated in the sediments, and where anoxic conditions may prevail. Because sedimentation rates are much lower in these isolated regions than off the delta lobe itself, these organic-rich zones actually contribute little to the total sediment accumulation. They therefore do not appreciably influence the regional petroleum geochemistry.

Table 2.1. Depositional Environments and Their Effect on Preservation of Organic Material

Growth Environment of Organisms	Transport Mode To Depositional Environment	Depositional Environment	Examples	Conditions In Depositional Environment	Preservation of Organic Material	Type of Organic Material	Fossil Fuel Potential
Land	Vertical sedimentation	Peat/Coal swamp	Dismal swamp, Virginia	Low energy; little oxygen; little bacterial activity; little admixed mineral matter	Quality: unoxidized Quantity: very high	Woody	Coal (perhaps some oil)
Land	Vertical sedimentation	Eutrophic lakes	Mountain lakes; Minnesota lakes	Low energy; some oxygen; moderate anaerobic bacterial activity; low to moderate admixed mineral matter	Quality: relatively unoxidized Quantity: high	Woody and Algal	Probably won't be preserved over geologic time
Land	Fluvial	Coastal or deltaic	Gulf Coast	High energy; much oxygen; high aerobic bacterial activity; high admixed mineral matter	Quality: highly oxidized Quantity: low	Woody	Gas and some oil
Lacustrine	Vertical sedimentation	Stratified lake	Green River Shale	Low energy; little oxygen; little bacterial activity; water stratification; moderate admixed mineral matter	Quality: unoxidized Quantity: high to very high	Algal	Oil

Table 2.1 (continued)

Marine	Coastal currents	Coastal marsh and embayments	Gulf Coast	High energy; much oxygen; high aerobic bacterial activity; high admixed mineral water	Quality: highly reworked by bacteria Quantity: low to moderate	Algal	Oil (quantity depends on rock thickness)
Marine-Brackish	Vertical sedimentation	Quiet deep bays or non-circulating waters	Black Sea	Relatively shallow; low energy; water stratification; little oxygen; low anaerobic bacterial activity; low admixed mineral matter	Quality: unoxidized Quantity: high	Algal and woody	Oil
Marine	Vertical sedimentation	Continental shelf, near upwellings within oxygen minimum layer	Phosphoria Fm.; Pacific Ocean west of Peru	200- to 1500-m water depth; little oxygen, low bacterial activity; chemical precipitation of minerals	Quality: unoxidized Quantity: high	Algal	Oil
Marine	Vertical sedimentation	Open ocean, abyssal depths	DSDP Leg 58 (North Philippine Sea)	Low energy; high oxygen from bottom currents; high aerobic bacterial activity	Quality: highly oxidized Quantity: very low	Algal	Nothing

Because they do not contain large volumes of organic-rich rocks that have an obviously high oil-source potential, deltas like the Mississippi Delta in the Gulf Coast have long been somewhat of an enigma to organic geochemists. In recent years, however, geochemists, often at the prodding of Gulf Coast geologists, have come to realize that a thick sequence of rocks containing only moderate proportions of organic matter can generate as much oil as thinner sequences of richer rocks. Because deltaic sequences are often very thick, large quantities of hydrocarbons can be generated in them.

The Gulf Coast, with its complex growth-fault and salt dome–diapir tectonics, also provides unusually good opportunities for petroleum migration and accumulation (see Chapter 5). Furthermore, deltaic environments like the Gulf Coast generally have excellent source rock–reservoir relationships because of lateral facies changes. Because migration and trapping efficiencies are very important in making the Gulf Coast a bountiful hydrocarbon province, it is necessary to be able to take these factors into account when making predictions of the volume of oil likely to be present in such an area. This requirement in turn points out the need for a totally integrated approach to oil exploration, in which both source rock evaluation and migrational efficiency are considered.

The Black Sea is an example of a basin which is permanently anoxic because of stratification of the water column and lack of mixing of shallow and deep waters. It was studied extensively on Leg 42B of the Deep Sea Drilling Project (Hunt and Whelan, 1978, and many other authors in the same volume). The restricted influx of both fresh and saline waters suppresses both aerobic and anaerobic bacterial activity and leads to the preservation of much of the organic material. The presence of the organic-rich sediment in turn insures that the bottom water will remain anoxic.

The Green River Shale was deposited under conditions which bear some resemblance to those prevailing today in the Black Sea. Both the Black Sea sediments and the Green River Formation were deposited under conditions of varying salinity as the bodies of water became progressively more isolated. Organic productivity in both situations was very high, and stratification of the water layers caused permanent anoxia at the sediment–water interface. Bacterial degradation was minimal, because under the conditions prevailing in most natural environments, anaerobic degradation is slow and inefficient compared to aerobic degradation.

One important difference from the point of view of petroleum geochemistry arises from the different sources of the organic material. The kerogen of the Green River Formation is thought to be composed largely of algal material, whereas the Black Sea sapropel contains predominantly terrestrial organic debris, with some admixed algal material. The oil-generative capacities of kerogens formed in the two environments will therefore probably be substantially different.

A somewhat rare type of depositional environment is that in which phosphatic shales form. These deposits often are linked to upwellings of nutrient-rich subsurface waters. Such waters support prolific populations of phytoplankton and their predators. A modern example is the fishing grounds on the continental shelf off Peru, and an ancient one is the Phosphoria formation of the midcontinental United States. The sedimentary rocks derived from phosphate-rich sediments are called *phosphorites*; good studies of them have been compiled by Manheim et al. (1975), Claypool et al. (1978), Powell et al. (1975), and Piper and Codispoti (1975). Phosphatic shale beds provide near-ideal conditions for the preservation of good oil-source material, viz., large quantities of algal bodies, short vertical transport to the bottom, and anoxic conditions at the sediment–water interface and in the sediments themselves.

In contrast to the phosphate-rich sediments discussed above, pelagic sediments usually have little potential for producing fossil fuel. Biological productivity in the open ocean is quite low because only small amounts of nutrients are available there. The ecologies in the open ocean are adapted to extracting the maximum nutritional value from whatever organic material is available, so any organic material which does reach the bottom will be highly reworked (oxidized). Finally, the deep bottom currents are generally oxidizing, and they therefore promote further destruction of the organic detritus. Most abyssal sediments contain extremely small amounts of organic carbon, and what organic material there is does not appear to be capable of generating gas or oil (Waples and Sloan, 1980).

In summary, we can make the following statements about preservation of organic material in sediments.

1. Preservation is an unusual event; most carbon is immediately recycled.
2. The degree of preservation varies from nearly 0% to almost 100%, depending upon many factors.
3. Microbial and nonbiological oxidation destroy most organic material. Significant amounts of preservation occur only where microbial activity is somehow reduced.
4. Phenolic compounds act as bacteriocides, especially in forested areas.
5. Lack of dissolved oxygen in water prevents aerobic bacterial activity. Anaerobic bacterial activity can occur under these conditions, but is generally much more selective and less thorough than aerobic activity.
6. Formation of geopolymers begins immediately after the death of the organism. The original geopolymers are gradually converted to kerogen, the composition of which changes throughout the sedimentary column.
7. Algal material is much more easily converted to petroleum than is terrestrial organic material.
8. High-energy environments are destructive of organic material, whereas such material may be preserved in low-energy environments.
9. To better understand the process of kerogen formation it will be necessary to develop a clearer understanding of the role of microorganisms in geochemical processes.

A long discussion of organic material in marine environments is given by Bordovskiy (1965). More recently, Demaison and Moore (1980) have discussed in detail the geologic, oceanographic, and climatic conditions that lead to the deposition of anoxic, organic-rich sediments.

3 COMPOSITION OF KEROGENS, BITUMENS, PETROLEUMS, AND NATURAL GASES

1. Introduction

Before discussing the chemical composition of the various types of organic materials listed in the title of this chapter, it is useful to discuss briefly the genetic relationships among them. (Coal is considered here to be a special kind of kerogen, so the discussions of kerogen in this chapter will implicitly include coal.)

Kerogen is usually defined as that organic material in sedimentary rocks which is insoluble in ordinary organic solvents. Because this is an operational definition, the exact quantity and chemical composition of kerogen in a given rock will depend upon many factors, such as the solvent used in extraction, the length of time used for the extraction, and the particle size to which the rock was ground before it was extracted. Significant problems can therefore arise when data from two different laboratories are compared. The insolubility of kerogen molecules derives from their large size, so it is molecular size more than anything else which distinguishes kerogen from the soluble portion of the organic material.

In unconsolidated sediments, the insoluble organic matter is not yet a true kerogen. Formation of geopolymers is still in its early stages, and the polymer molecules are small compared to kerogen molecules. These early geopolymers are often called humic acids and fulvic acids; they are discussed briefly in Chapter 2.

The organic material extracted from the rock with a solvent is called *bitumen*. Kerogen and bitumen together constitute the *total organic carbon* in any rock. Bitumen composition and quantity are obviously also dependent upon extraction conditions. Once extracted, some bitumens are viscous liquids, while others look more like dried scabs. Colors vary from pale yellow to dark brown.

Petroleum is liquid or solid organic material which can be obtained from wells or natural seepages. Varieties of petroleum include waxy solids, black liquids, and transparent liquids with the appearance of gasoline.

We may think of bitumen as dispersed petroleum. As they migrate out of the source rock, through a permeable conduit, and into a reservoir rock, the bitumen molecules coalesce to form high local concentrations of migrating organic fluid. At some point the organic molecules are sufficiently concentrated that we may refer to them as petroleum.

The distances required for movement before bitumen coalesces into petroleum must be extremely variable. In fractured shale reservoirs little or no movement is necessary, and the terms bitumen and petroleum are practically synonymous. In more traditional source-rock–reservoir relationships, significant movement may be involved. Migration induces changes in the chemical composition of the migrating fluid, but compositional differences between bitumens and petroleums are not the essence of the distinction between these two fluids.

Natural gas is of two types: wet and dry. Dry gas contains 90-100% methane (sometimes the range is given as 95-100%), with small amounts of ethane and higher hydrocarbons. Wet gas contains greater proportions of ethane, propane, butane, etc.

Some hydrocarbon mixtures which are gaseous at subsurface temperatures become liquid when cooled to surface temperature. These fluids therefore condense and liquify when produced, and consequently are called *condensates*. There is, however, no fundamental difference between wet gases and condensates. Condensates are composed mainly of compounds whose molecules contain at least five carbon atoms.

Natural gas can be formed in three ways. Some anaerobic bacteria produce methane as a metabolic by-product at shallow depths and low temperatures. This biogenic methane usually escapes rapidly into the air or the overlying water column through unconsolidated sediments. Evidence is mounting, however, that many deposits of natural gas are formed at least in part from biogenic methane.

Natural gas can also be produced directly from kerogen by thermal decomposition. The reactions which produce methane require more energy than do the reactions which produce bitumen, so gas is produced after bitumen is formed.

Gas can also be formed by thermal cracking of bitumen or petroleum in the subsurface. This transformation also requires a large amount of energy, and is a late-stage development. Most of the world's gas reserves probably are produced by *in-situ* cracking of petroleum.

There is also a class of material referred to variously as "solid bitumens," "solidified bitumens," "pyrobitumens," etc. These materials vary from being totally soluble in organic solvents to being almost totally insoluble. Their common feature is that they all appear to have once been normal bitumens or migrating petroleums which turned solid and stopped moving, probably because they lost their volatile and mobile components. It is useful to think of these materials as being genetically related to bitumen and petroleum.

2. Kerogen

Kerogens are formed primarily from the five elements carbon, hydrogen, oxygen, nitrogen, and sulfur, but the relative proportions of these elements vary over a wide range. Tissot and his co-workers have grouped kerogens into four types based on their carbon, hydrogen, and oxygen content. An example from each group is shown in Table 3.1, with a coal included for comparison. It is readily apparent that both the H/C and O/C ratios change significantly over the progression from Type I to Type IV kerogen. The change in hydrogen content is directly related to fundamental changes in the structure and chemistry of the kerogen. Kerogens which have high hydrogen contents (high H/C ratios) have a relatively high proportion of alkane chains and saturated rings (like cyclohexane) in their structures. Kerogens with low H/C ratios are more aromatic (contain more unsaturated rings). One cause of the differences in kerogen types is that they are formed from different types of organic source material. As was mentioned in Chapter 2, woody plants contain a lot of lignin, which decomposes to yield phenols. When large numbers of phenolic components combine in the kerogen structure, the kerogen will be aromatic. Woody kerogens (Type III), therefore, have low H/C ratios. Most coals would be classified as Type III kerogen.

Algal kerogens (Type I), in contrast, are hydrogen-rich, because phytoplankton do not contain lignin or cellulose. Such kerogens often contain high concentrations of alkanes and fatty acids, and are therefore rich in hydrogen. Algal kerogens usually have high H/C ratios.

Type II kerogens are formed from *lipid* (fat and wax) components, particularly the waxy outer coatings of pollen grains, spores, and leaves. They may also contain, or consist predominantly of, bacterial bodies (Lijmbach, 1975). As a consequence, they are also hydrogen-rich.

Woody kerogens contain more oxygen than do algal kerogens, because both lignin and cellulose are oxygen-rich. Both H/C and O/C ratios are thus useful in deducing the origin of a given kerogen. Neither nitrogen nor sulfur seems to have any fundamental structural significance, although it has been proposed that nitrogen content is dependent upon the type of source organic material (Waples, 1977).

All the kerogens shown in Table 3.1 are thermally immature; that is, they are in the diagenetic stage of transformation. The elemental compositions therefore reflect their sources very accurately. When catagenesis (thermal transformation) begins to occur, the elemental composition of each kerogen changes, in ways that are discussed in detail in Chapter 4. For the moment, however, consider Figure 3.1, which is a schematic plot of H/C ratio against O/C ratio. This type of diagram is called a *Tissot diagram*. It is based on the older van Krevelen diagrams from the coal literature. Each of the branches represents a different type of kerogen, as noted in the figure. Thermal maturity increases in the directions indicated by arrows. Type I and Type II kerogens, which have little oxygen, lose mostly hydrogen; Type III and Type IV kerogens lose mostly oxygen. The actual chemistry of these transformations is considered in Chapter 4.

As a kerogen undergoes catagenesis, it becomes increasingly difficult to say what it looked like when it was immature. Take as an extreme example point A in Figure 3.1. A kerogen with this composition could originally have been of Type I, II, III, or IV. Catagenesis thus complicates our use of elemental compositions of kerogen as indicators of the parent organic material.

The chemical composition of a kerogen is also reflected in its microscopic appearance. The combined efforts of palynologists, coal petrographers, and organic geochemists have produced a classification system for kerogen particles, or *macerals*, that is based on their visual appearance under a transmitted light microscope. The classification system, which is a simplification of the older system applied in coal petrology, recognizes four major types of kerogen. These four types, which correspond to Tissot's Types I–IV, are called alginite, exinite, vitrinite, and inertinite. Exinite is sometimes called liptinite; in other cases alginite and exinite occasionally are grouped together and called liptinite.

Alginite includes material of algal origin. *Exinite* includes spores, pollen grains, and bits of leaf-cuticle wax. *Vitrinite* is identical to the dominant maceral of humic coal, and bears the

Table 3.1. Elemental Composition of Type I, II, III, and IV Kerogens and a Lignite[a]

Kerogen Type	Basin	Age	Weight %					Atomic Ratios	
			C	H	O	N	S	H/C	O/C
I	Uinta, Utah, USA	Green River Shale, Eocene	75.9	9.1	8.4	3.9	2.6	1.44	0.08
II	Paris, France	L. Toarcian	72.6	7.9	12.4	2.1	4.9	1.30	0.13
III	Douala, Cameroun	U. Cretaceous	72.7	6.0	19.0	2.3	0.0	0.99	0.19
IV	—	—	88.0	4.2	6.7	0.6	0.5	0.57	0.06
Coal	Yugoslavia	Pliocene	68.6	5.1	21.2	2.3	2.5	0.90	0.23

[a] All kerogens and coals are in the diagenetic stage of transformation. After **Tissot** and **Welte** (1978) and Dormans et al. (1959).

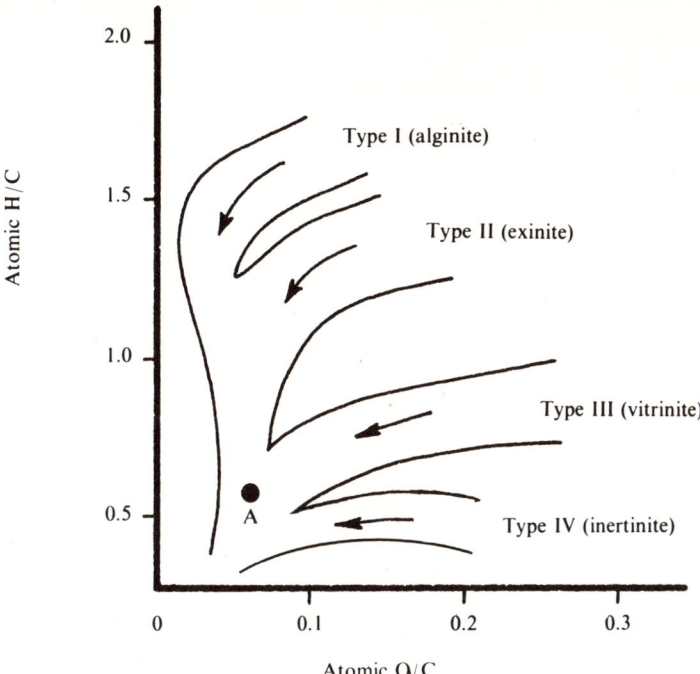

Figure 3.1. H/C versus O/C plot for kerogens of Types I, II, III, and IV. Maturity increases along each pathway as it leads down and to the left.

same name. It consists largely of woody and cellulosic debris. *Inertinite* is thought to be material of various origins which has undergone extensive oxidation prior to deposition. It may have been oxidized in forest fires, by bacterial or air oxidation, or may in some cases represent organic material from a previous depositional cycle which was eroded and reworked. Figure 3.1 shows the four maceral types on a Tissot diagram.

Much work has been done on the specific chemical structures of certain kerogens, especially the Green River Shale, but these are of interest mainly to the specialist. The recent volume by Tissot and Welte (1978) contains references to many such studies. A good understanding of the structural significance of H/C ratios, of the correlation between maceral type and chemical composition, and of the effects of catagenesis on chemical composition is sufficient for exploration geologists. Catagenetic transformations of kerogen are discussed in Chapter 4.

3. Petroleum and Bitumen

Like kerogens, petroleums exhibit a very wide range of chemical compositions. Because petroleum molecules are small and can therefore be handled chemically with relative ease, we are able to separate several distinct groups of compounds according to their chemical properties. The four major classes of compounds present in bitumens and petroleums are saturated hydrocarbons, aromatic hydrocarbons, resins, and asphaltenes.

Saturated hydrocarbons are defined in Chapter 1; they include compounds such as those shown below.

Aromatic hydrocarbons, also discussed in Chapter 1, include all hydrocarbons that have at least one aromatic ring.

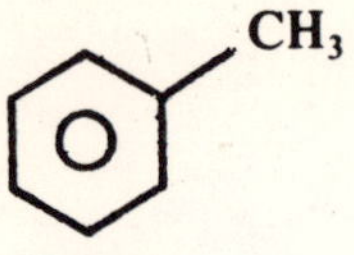

Toluene

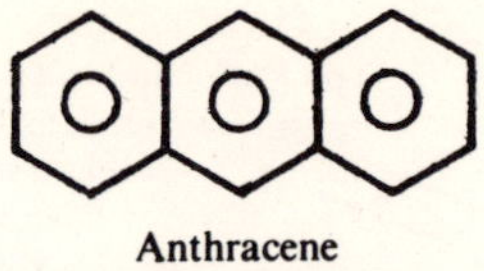

Anthracene

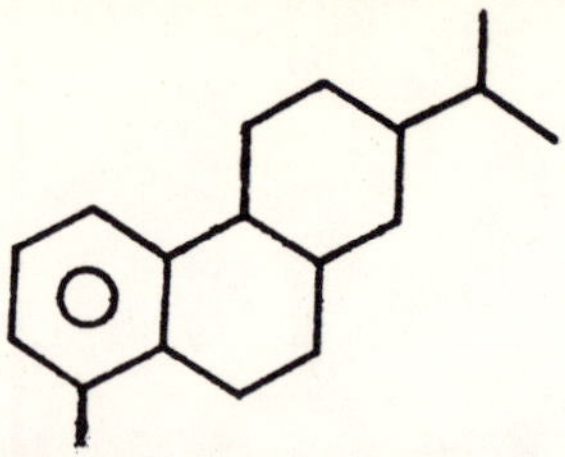

Molecules which contain one or more heteroatom are often referred to as *resins*, or "polars." There are many kinds of molecules represented in this group, most of which occur only in minute quantities in petroleums and bitumens. Some examples are shown below.

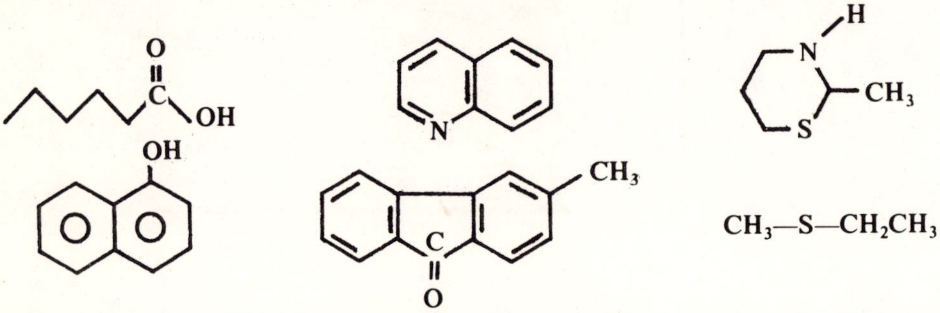

$CH_3-S-CH_2CH_3$

Examples of Resin Structures

Porphyrins (see Chapter 1) also belong to the resin group. The resins are the most polar molecules found in petroleum and thus interact most strongly with water and clay mineral surfaces. They are therefore very important in determining the mobility of crude oil.

Asphaltenes are high-molecular-weight compounds that contain significant numbers of heteroatoms and have a generally aromatic structure. An example is shown below. When rings are joined along a side, they are said to be *fused,* or *condensed.*

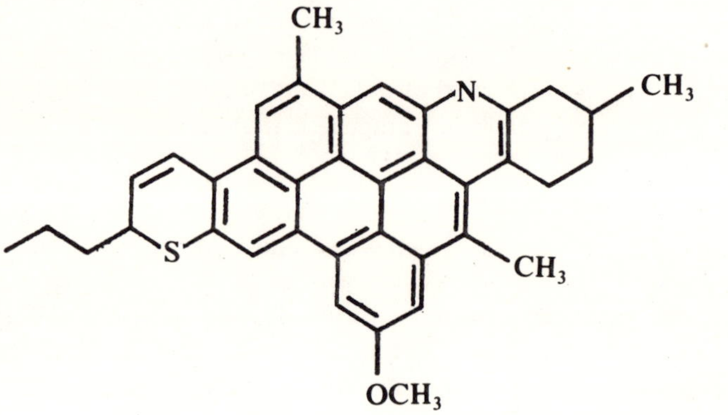

Example of Asphaltene Structure

Because the aromatic sheets present in asphaltenes are flat, or nearly flat, asphaltene molecules sometimes aggregate in stacks, like a sheaf of papers. These aggregates tend to behave as a unit, forming particles with effective molecular weights of 50,000 or more. Yen (1972) has studied asphaltene structures in some detail.

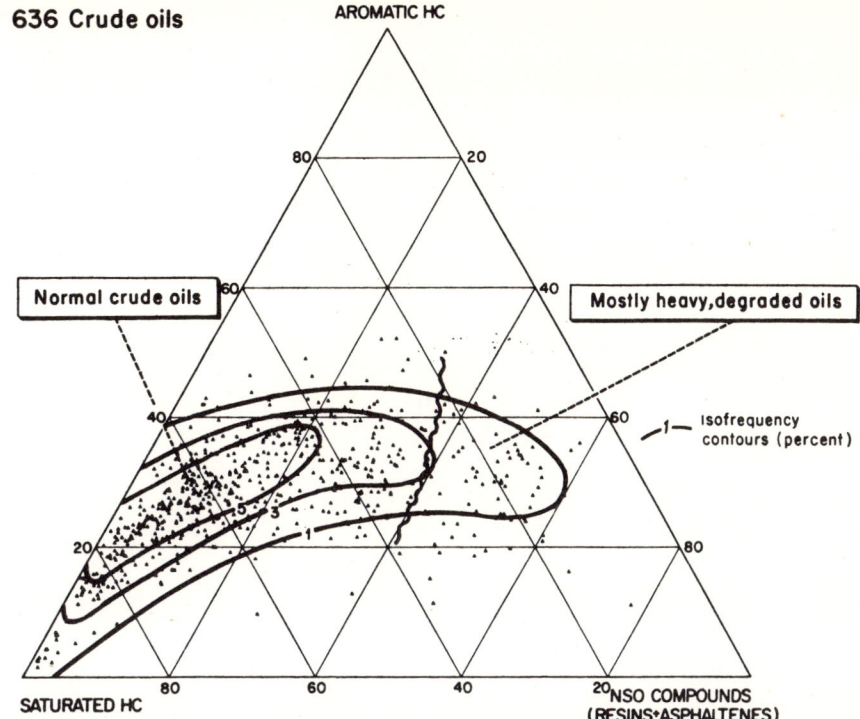

Figure 3.2. Ternary diagram showing the gross composition of 636 crude oils. (From Tissot and Welte, 1978; republished with permission of Springer-Verlag)

The large size of asphaltene aggregates makes them highly insoluble, even in petroleum. If a light hydrocarbon cosolvent such as methane or pentane is added to crude oil, the asphaltenes precipitate because the light hydrocarbons are poor solvents for the asphaltenes. This process is used both industrially and in the laboratory to remove the asphaltene fraction. It also undoubtedly occurs in nature in petroleum reservoirs, leading to a lightening of the crude oil and possibly plugging reservoir pores.

If one combines the resins and asphaltenes into a single group, usually designated as *NSO* (because they contain nitrogen, sulfur, and oxygen), one can show the gross composition of a crude oil by means of a triangular diagram such as that illustrated in Figure 3.2. It is apparent that crude oils come in a wide range of compositions. Average values compiled by the French Petroleum Institute for 517 normal producible crude oils are given in Table 3.2. The component which varies over the widest range is the saturated hydrocarbons, mainly because these components are susceptible to bacterial destruction in the reservoir.

The composition of a crude oil depends on several factors, viz., the type of organic source material, oxidation and other diagenetic transformations in the depositional environment, the nature of migration between source rock and reservoir, and reservoir transformations, including water washing, bacterial degradation, and thermal alterations. Let us explore some of these factors briefly.

As was noted in Chapter 2, algal material tends to yield predominantly saturated hydrocarbons, especially cyclic compounds, whereas a woody-plant contribution will increase the aromaticity of an oil. Interestingly, however, the most waxy crudes, which contain large amounts of long-chain *n*-paraffins, seem to be derived from terrestrial plant sources. This fact probably reflects a strong contribution from leaf and other plant waxes, which are rich in long-chain hydrocarbons and fatty acids.

Table 3.2. Average Gross Composition of 517 Normal Producible Crude Oils

Saturated hydrocarbons	57.2%
Aromatic hydrocarbons	28.6%
NSO's	14.2%

Bacterial and nonbiological oxidation in the depositional environment results in many changes, one of which is the incorporation of sulfur into the organic material. Bacterial reduction of inorganic sulfate from sea water produces sulfide ion and elemental sulfur. In clastic sediments there is an abundance of heavy-metal ions, which under basic conditions can form insoluble sulfides. The most common of these are marcasite and pyrite, FeS_2. Sulfur is thus kept out of kerogen and bitumen. Carbonates, on the other hand, lack the heavy-metal ions; the bacterially-produced sulfur in such sediments eventually reacts with organic material, yielding high-sulfur kerogens and bitumens.

Migration apparently produces important changes in composition, because disseminated bitumens differ greatly from reservoired petroleums in their chemical composition. There is clear evidence that the heavier and more polar components of petroleums are selectively lost during migration, resulting in a marked decrease in the concentration of NSO's between source rock and reservoir. The relative concentration of aromatic hydrocarbons also seems to decrease. Crude oils become progressively lighter and more *paraffinic* with increasing distance of migration, but the fact that other factors are operative simultaneously with migration makes these changes difficult to document.

Water washing in the reservoir gradually removes those petroleum constituents which are most soluble in water. Thus the lightest hydrocarbons, such as methane, ethane, and propane, aromatic compounds like benzene and toluene, and small heterocompounds like phenols and fatty acids are all selectively removed. Depending upon whether the original oil had more light alkanes or light aromatics, the washed oil could become either more paraffinic or more aromatic.

Bacterial degradation in the reservoir preferentially removes *n*-alkanes, so bacterially degraded oils will become more aromatic. In extreme cases, the *n*-alkanes and isoprenoids may be absent altogether. See Bailey et al. (1973) for a more detailed discussion of biodegradation and water washing.

Thermal transformations in reservoirs have been studied in some detail by Rogers et al. (1974) and Milner et al. (1977). Cracking reactions, in which large molecules are converted into smaller ones, are common. One such reaction is given in equation 3.1.

$$CH_3(CH_2)_6CH_3 + H_2 \xrightarrow{heat} 2\ CH_3CH_2CH_2CH_3 \qquad (3.1)$$
$$\text{\textit{n}-Octane} \qquad\qquad\qquad \text{\textit{n}-Butane}$$

$$\text{Cyclohexane} \xrightarrow{heat} \text{Benzene} + 3H_2 \qquad (3.2)$$

As the equation indicates, cracking an alkane to produce two smaller alkanes requires that hydrogen be added. Molecular hydrogen (H_2) is not available in an oil pool, so the source of the hydrogen is probably other organic molecules. As equation 3.2 shows, cyclic compounds can readily give up hydrogen and become aromatics, which are unusually stable.

Combining equations 3.1 and 3.2 yields equation 3.3. Such hydrogen-transfer reactions

$$\text{C}_6\text{H}_{12} + 3\text{CH}_3(\text{CH}_2)_6\text{CH}_3 \longrightarrow \text{C}_6\text{H}_6 + 6\,\text{CH}_3\text{CH}_2\text{CH}_2\text{CH}_3 \tag{3.3}$$

are called *disproportionation* reactions. Very often the aromatic products are asphaltenes. As disproportionation proceeds, the transformed oil becomes enriched in both the very light alkanes and in the very large asphaltenes. Eventually the solubility of the asphaltenes will be exceeded, and they will begin to precipitate, forming reservoir bitumens and tars. The overall producing quality of the oil is improved by disproportionation, but plugging of pores by asphaltenes can be a problem.

Of the four classes of crude oil components, the saturated hydrocarbons have been by far the best studied and most useful for petroleum geochemistry. The major components of the saturated hydrocarbon fraction are *n*-alkanes, branched alkanes, and cycloalkanes (*naphthenes*).

All of the *n*-alkanes from C_1 through C_{40} have been identified in many crudes. They usually represent approximately 15–20% of the total crude, but they may be absent in highly degraded oils. Waxy crudes may contain higher concentrations of *n*-alkanes. The molecular weight distributions of *n*-alkanes in petroleum usually have a single mode, with the maximum occurring at any value from C_5 to C_{31}. Examples of distribution curves for six different crude oils are shown in Figure 3.3.

It is clear from a glance at Figure 3.3 that the distributions are not smooth. Some oils, like North Smyer, contain mainly the very short *n*-alkanes. These oils have suffered much cracking, and are therefore very mature. Others, such as State Line and Uinta Basin, contain large amounts of heavy *n*-paraffins. Most of the curves in Figure 3.3 show at least some sawtooth nature. The odd-carbon *n*-alkanes are present in larger quantities than the adjacent even-

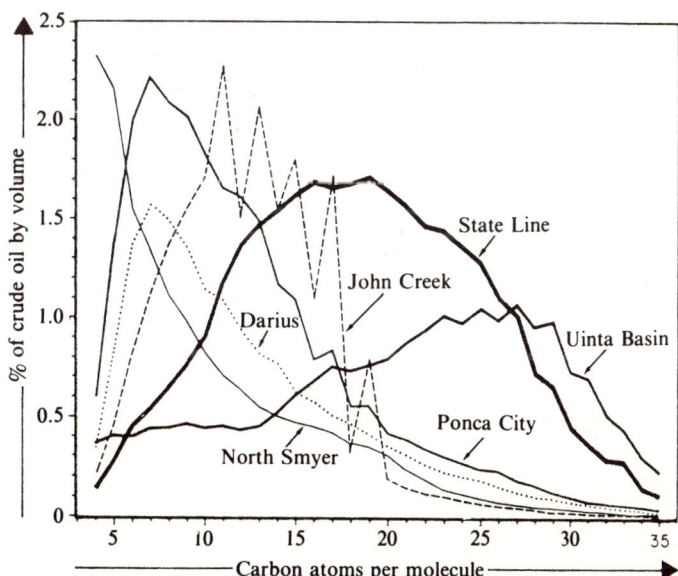

Figure 3.3. Distribution of *n*-alkanes in different types of crude oils. (From Martin et al., 1963; republished with permission of Nature)

carbon compounds. This is called an "odd-carbon preference." It is a common feature of many oils and bitumens, and is related to the nature of the original organic material. It is known that in land plants the odd-carbon n-alkanes predominate over the even-carbon ones by a factor of five or so. This predominance is maintained as the dead plant is buried in sediment, but is gradually lost as chemical reactions transform the organic material. Thus the degree of odd-carbon preference is indicative of both the amount of terrestrial contribution and the degree of thermal maturity of the oil. This phenomenon is discussed further in Chapter 4.

The *isoprenoids* are the most useful and interesting group of branched hydrocarbons. The structures of several geochemically important isoprenoids are given in Table 3.3. Phytane, pristane, and the lower isoprenoids are probably derived from phytol, which is a portion of the chlorophyll molecule (see Figures 1:1 and 1.2). The C_{17} isoprenoid is usually absent, or present in very small concentrations, for reasons discussed by McCarthy and Calvin (1967). These isoprenoids are called "regular" because they have a methyl branch on every fourth carbon atom. Regular isoprenoids with more than 20 carbon atoms have been found in petroleums and bitumens many times, but their origin and significance are unclear because it is unlikely that they are derived from phytol.

Table 3.3 shows three important nonregular isoprenoids. Lycopane, for example, can be thought of as two phytane molecules joined asymmetrically, and squalane could be derived from farnesane in the same way. This is how these carbon skeletons are formed in living organisms, but those syntheses occur enzymatically. Squalane, lycopane, and perhydro-β-carotene thus

Table 3.3. Structures of Geochemically Important Isoprenoids

Structure	Name	Number of Carbon Atoms
	Phytane	20
	Pristane	19
	Norpristane	18
	—	17
	—	16
	Farnesane	15
	Squalane	30
	Lycopane	40
	Perhydro-β-carotene (β-carotane)	40

probably come from the unsaturated, naturally occurring hydrocarbons squalene, lycopene, and β-carotene, respectively. These compounds can serve as important geochemical indicators and as fingerprints relating crude oils to their source rocks.

Cyclic compounds represent the most complex group of saturated hydrocarbons. The small members of this group, the cyclohexanes and cyclopentanes, are common constituents of petroleum, but are not of great geochemical usefulness. The most complex members, the steranes and triterpanes (Figure 1.1), in contrast, convey a great deal of important geochemical and paleoecological information. They will be discussed in more detail in Chapters 4 and 7.

Aromatic hydrocarbons are usually classified according to the number of aromatic rings they contain. Naphthenoaromatics are compounds which contain both saturated and aromatic rings. Two examples are shown below.

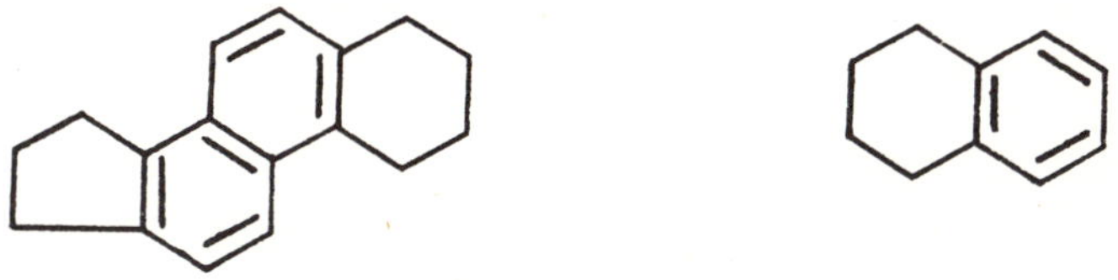

Examples of Naphthenoaromatics

Detailed analyses of groups of such compounds can be carried out by mass spectroscopy, but the technique is not in general use in exploration geochemistry.

The group of NSO compounds of most importance to petroleum geochemistry is the porphyrins. The basic porphyrin nucleus (shown below) can be modified in many ways to give a large number of distinct porphyrin molecules. Chlorophyll, the precursor of the porphyrins found in sediments, contains a magnesium (Mg^{2+}) ion, and the most important and obvious change is in the nature of the metal ion (M^{2+}) *chelated* in the center of the molecule. Nickel and vanadyl are the ions most commonly present in sedimentary porphyrins. A few copper porphyrins are also known.

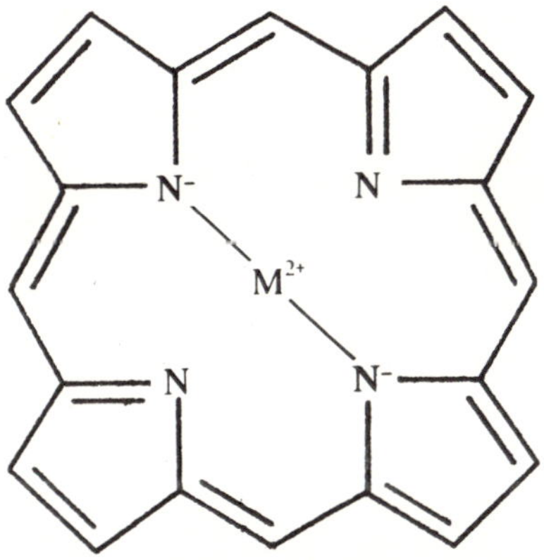

The porphyrin nucleus

Each of the positions around the outside of the porphyrin molecule can be substituted by various alkyl groups (methyl, ethyl, etc.). A large number of possible permutations of the number, size, and positions of alkyl substituents gives rise to a whole family of porphyrins (Baker, 1966; Didyk et al., 1975). Porphyrin concentrations may reach 400 ppm in some crude oils, but they are generally much lower than that. Much of the metal present in petroleum is found in porphyrin complexes.

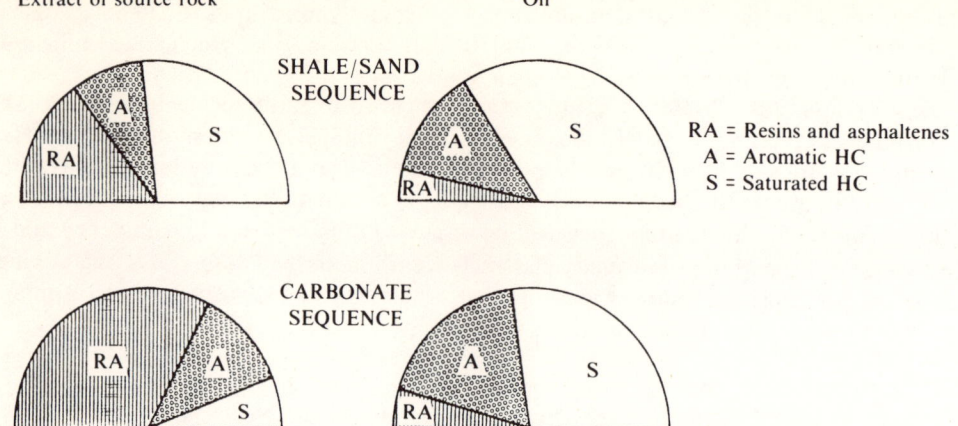

Figure 3.4. Comparison of gross chemical composition of crude oils and source rock bitumens in a shale/sand and carbonate sequence. (From Tissot and Pelet, 1971; republished with permission of Applied Science Publishers, Ltd.)

Bitumen, the soluble material found in sedimentary rocks, contains the same groups of compounds as petroleum, but in different proportions. Figure 3.4 illustrates these gross compositional differences for both shale and carbonate source rocks. Saturated and aromatic hydrocarbons play a lesser role in bitumen than they do in petroleum, while resins and asphaltenes are much more important in bitumens. Nitrogen, oxygen, and sulfur contents are significantly greater in bitumens than in petroleums.

Within each class of compounds, the heavier components often are more abundant in bitumen than they are in petroleum. For example, a bitumen whose *n*-alkane distribution curve (Figure 3.3) has a maximum near C_{30} might, after migration, exhibit an *n*-alkane maximum somewhat below C_{30}. Examples are also known in which a bitumen and its related petroleum have identical *n*-alkane distributions. Among polycyclic saturated and aromatic hydrocarbons, the average number of rings per molecule will decrease. All these changes are related mainly to migration in complex ways that are discussed in Chapter 7.

4. Solidified Bitumens

These materials have a wide range of compositions. They are not of direct interest in petroleum exploration, but do have some implications for migrational theories and are mentioned in Chapter 5. The reader is referred to King et al. (1963) and Hunt (1963) for more detailed studies.

5. Natural Gas

The composition of natural gas is dependent upon both the type of organic material from which it was formed, and the time in the diagenetic-catagenetic sequence when it was formed.

Biogenic gas is formed by bacterial activity in the early stages of diagenesis. The only hydrocarbon produced in measurable quantity is methane, but trace amounts of C_2–C_6 hydrocarbons may also be present. Carbon dioxide (CO_2) and hydrogen sulfide (H_2S) are also produced. Both CO_2 and H_2S are often removed to a large extent by precipitation as $CaCO_3$ and FeS_2. During chemical diagenesis, some CO_2 is formed by cleavage of carboxyl (COOH) groups, particularly those derived from terrestrial organic material. Some hydrocarbons may also be formed by chemical reactions, but not in large amounts.

During the early stages of catagenesis, little gas is produced. Most of the cracking reactions produce the larger bitumen molecules, because the activation energies required for bitumen formation are lower than those for gas formation. Any gas that is formed will contain significant amounts of the C_2–C_6 components and will be a wet gas or a condensate.

During the late stages of catagenesis and the following high-temperature stage called *metagenesis*, when the kerogen's bitumen-forming possibilities have been exhausted, methane is the dominant product. Late catagenetic and metagenetic gases are dry gases. H_2S can also be formed at these temperatures, especially in carbonate sequences. Natural gas also contains helium and argon from radioactive decay processes, and nitrogen, which is thought to be of volcanic origin.

Carbon isotope ratios (see Chapters 6 and 7) are extremely useful in classifying gases as biogenic or thermal. Biogenic methanes are greatly depleted in ^{13}C compared to thermal methanes. A combination of carbon isotope data and a "wetness" parameter like C_1/C_2+ can easily distinguish among biogenic, catagenetic, and metagenetic hydrocarbon gases (Bernard et al., 1976; Sackett, 1977).

Natural gas can exist alone in reservoirs, it can be dissolved under pressure in petroleum, it can overlie petroleum as a gas cap, or it can form solid crystalline hydrates where temperatures never exceed 30 °C. The actual temperature-stability range for methane hydrates depends greatly on pressure, and they generally form only beneath strata where temperatures never exceed 5 °C. *Methane hydrates* are found in the arctic in regions of permafrost and can also exist beneath the sea floor (McIver, 1974; White, 1979; Tucholke et al., 1977). Gas hydrates are members of the structural class of compounds called clathrates—they consist of water molecules arranged in a rigid cage-like structure, each cage containing a single gas molecule. Methane hydrates are the most common, but ethane, propane, and butane can also form stable crystals.

Methane concentrations in hydrates are about 60 times higher than the maximum concentrations which can be attained in water solution, so hydrates are a very efficient means of storing methane. Since they form at shallow depths, they represent a possible trapping mechanism for biogenic gas. An article by Hitchon (1974) gives a good overview of gas hydrates.

The chemical compositions of kerogens, bitumens, petroleums, and natural gases are determined both by the type of source organic material and by the transformations which this material has undergone in the geosphere. Rather than memorizing compositional data, it is desirable to learn a few principles which govern initial composition, and a few more which govern the transformation processes. An understanding of these principles will enable one to make intelligent interpretations of the geological processes which have created the organic material encountered during exploration activity.

4 | CATAGENESIS OF ORGANIC MATERIAL AND THE FORMATION OF OIL AND GAS

1. Introduction

The chemical reactions which result in the transformation of kerogen into bitumen, and which ultimately yield petroleum and natural gas, require relatively high temperatures in order to occur even within the vast periods of time available. These thermally-induced reactions are called catagenesis. To understand the causes and results of catagenesis, it is necessary to study three aspects of the process, *viz.*, the factors which influence the rate of catagenesis, the effect of catagenesis on kerogen composition and structure, and the mobile products (bitumen and gases) of catagenesis.

2. Kinetics of Catagenesis

Kerogen catagenesis is very similar to other cracking reactions in that the large kerogen molecules decompose thermally to yield the smaller molecules called bitumen. Like other cracking reactions, kerogen decomposition can occur at a variety of sites in the molecule. Figure 4.1 illustrates two possible cleavages. If a carbon–oxygen bond breaks, a particular bitumen molecule is produced. If a carbon–carbon bond breaks, a different molecule of bitumen is released. There are in fact many other possible ways in which this (or any) kerogen molecule can break down. Each different pathway represents a slightly different chemical reaction, and each yields a slightly different bitumen molecule.

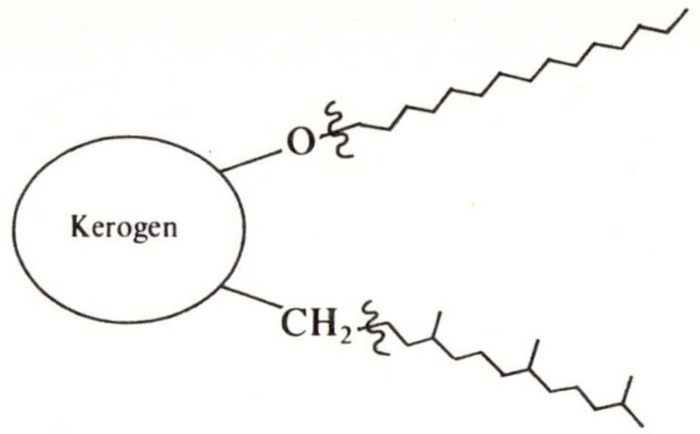

Figure 4.1 Schematic view of a kerogen molecule, showing two possible cleavage modes. The upper one involves breakage of a C–O bond, which requires relatively little energy. The lower one, in which a C–C bond is broken, requires more energy.

Not all these chemical reactions occur with the same ease. A bond between a carbon atom and a heteroatom such as oxygen, nitrogen, or sulfur generally will break more easily than will a carbon–carbon bond. Single bonds break more easily than double bonds.

Kerogen decomposition is a thermal process and the barrier that a molecule must surmount before decomposition occurs is called the *activation energy*. If the molecule is to pass over this energy barrier and decompose, energy must be put into it. This is why deep burial of sediments is vital to catagenesis. Temperature increases downward through the earth's crust at a rate called the *geothermal gradient*. As the depth at which a sediment is buried increases, its temperature also increases, and chemical reactions within the kerogen are facilitated. Connan (1974), Lopatin (1971), and Waples (1980) have all shown that the rate of catagenesis approximately doubles when the temperature increases by 10 °C. These observations are consistent with the predictions of chemical kinetics—specifically, that the decomposition of kerogen is a first-order process.

Given the relationship between time and temperature as they influence the rate of catagenesis, geologic time can be substituted for temperature in the catagenetic process. For example, if a million years were required to accomplish a certain transformation at 100 °C, the same transformation could be carried out at 90 °C in 2 million years, or at 80 °C in 4 million years. The interplay of time and temperature in catagenesis is dealt with in much more detail in Chapter 8.

Simoneit et al. (1979) have shown that generation of light hydrocarbons can occur even in unconsolidated sediments as a result of igneous intrusions. The high temperatures reached allow catagenesis, and even metagenesis, to occur in very short periods of time. Because the thermal effects of such intrusives are quite localized, however, it is unlikely that intrusions have played a significant role in oil generation. Plutons, on the other hand, may be important (e.g., the Baltimore Dome) because of their greater heating capacity.

Catalysts can increase the rate of chemical reactions by providing alternate pathways that have lower activation energies. Many mineral surfaces seem to have some catalytic effect on kerogen catagenesis (Shimoyama and Johns, 1971, 1972; Jurg and Eisma, 1964). Questions about whether mineral catalysis actually is important in subsurface kerogen catagenesis, and whether clay minerals or carbonates are more effective catalysts have not yet been answered, but it seems likely that the role of catalysts is not uniform in all sediments. Many workers believe that clays probably are more effective catalysts, and that catagenesis proceeds more rapidly in shales than it does in carbonate rocks. Others are not convinced that mineral catalysis is important in sediments.

The above analysis leads to two important conclusions. Firstly, the bitumen produced at different times during catagenesis of kerogen will have a chemical composition that depends upon the particular cleavage reactions which predominated at that time. Secondly, kerogens of different compositions will react at different rates. A kerogen with many C–O bonds, for example, will undergo catagenesis more rapidly than a kerogen whose molecules contain mostly C–C bonds. A corollary of this is that some kerogens will undergo catagenesis at lower temperatures (or in shorter lengths of time) than others.

In its early stages, catagenesis results in a net transformation of kerogen into bitumen. The quantity of kerogen in the sediment gradually decreases, and the amount of bitumen increases. If the temperature of the sediment is continually increasing because of ongoing subsidence, the rate of bitumen formation will increase exponentially. Eventually, however, the rate of catagenesis will decrease significantly because possible cleavage sites in the kerogen become scarce. This leads to a net decrease in the rate of bitumen formation, even though temperature is still increasing. Eventually, bitumen formation stops altogether. This sequence of events is shown in Figure 4.2.

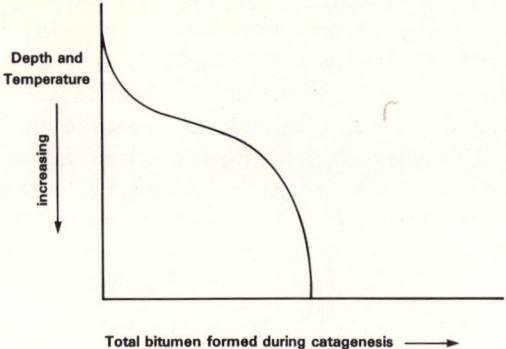

Figure 4.2. Generalized plot of the dependence of bitumen generation in a source rock on the degree of thermal maturation of the rock.

This analysis, however, does not tell the whole story of the bitumen content of rocks. For one thing, some diagenetic (inherited) bitumen is present even before catagenesis becomes significant. For another, not all the bitumen formed during catagenesis remains in the source rock.

During the early stages of catagenesis, bitumen loss is minimal, but during the late stages, when bitumen formation has decreased in importance, migration and destruction will dominate. These effects lead to a net decrease in total bitumen content of the rock in the late stages of catagenesis, as shown in Figure 4.3. Note that Figure 4.2 represents "total bitumen formed," while Figure 4.3 shows "total bitumen present." The differences between the two curves stem from the contribution from diagenetic bitumen and from migration/destruction effects.

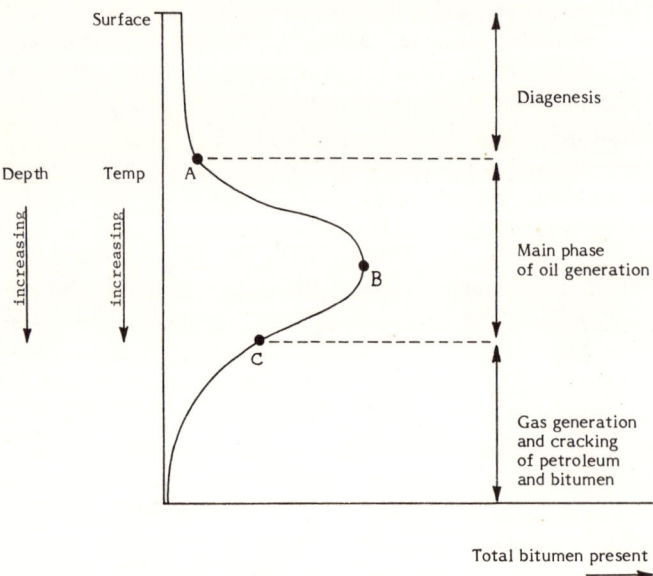

Figure 4.3. Generalized plot of the bitumen concentration in a source rock as a function of thermal maturation of the source rock.

The part of the section from the surface to point *A* in Figure 4.3 is the zone of diagenesis. Here the rate of thermal reactions is too slow to be of significance. The small amount of bitumen present in these sediments is inherited directly from the original source material, and is called diagenetic, or "inherited," bitumen. From *A* to *B*, there is a strong net production of bitumen, with relatively little destruction or migration. From *B* to *C*, bitumen formation gradually diminishes, and is now subordinate to the migrational or thermal destruction processes which remove it. Below point *C*, bitumen formation has essentially ceased and gas formation and bitumen migration are dominant.

This type of curve appears to describe adequately all examples which have been studied. The depth at which catagenesis begins depends upon the age and temperature of the sediments, as Figure 4.4 illustrates. The four basins described in Figure 4.4 are of very different ages, and they have different geothermal gradients. Nevertheless, the general shapes of the curves in Figure 4.4 all conform to the model. Note that the total amount of bitumen formed depends upon kerogen type, with Type III kerogen being significantly less productive than Types I and II. Type IV produces essentially no bitumen.

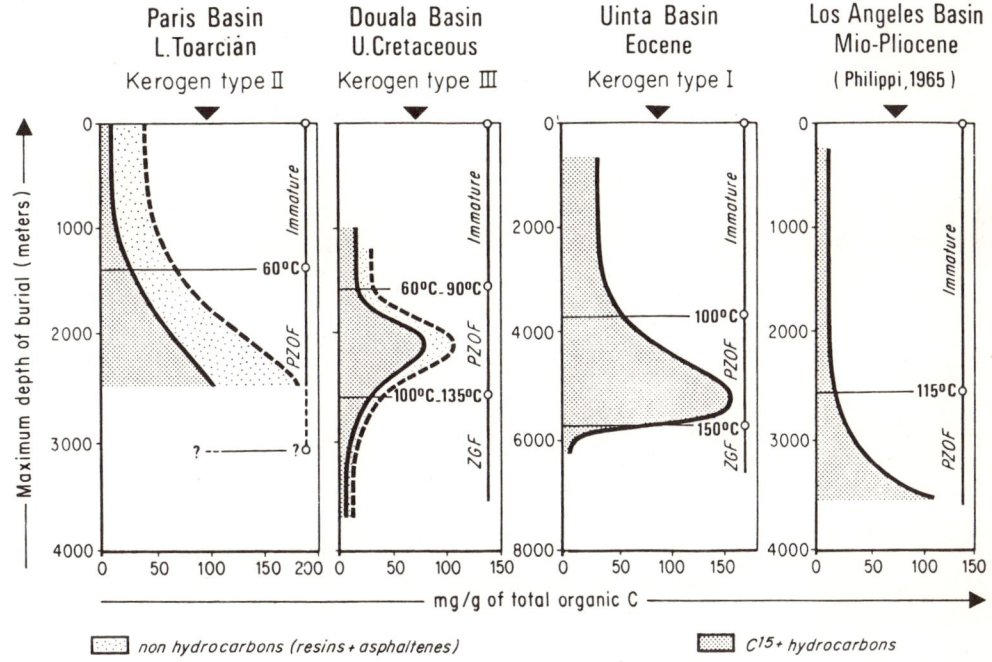

Figure 4.4. Formation of a bitumen as a function of burial depth in different basins. Corresponding present-day temperatures are shown. In the Doula Basin, the first temperature is the present one, and the second one is a calculated paleotemperature derived by Tissot and Espitalié (1975). (After Tissot and Welte, 1978; republished with permission of Springer-Verlag)

Total bitumen content has been used with some success to define the oil-generative zone in well profiles (see Figures 4.3 and 4.4). It should be noted, however, that no sense can be made of a single bitumen value; the entire profile must be constructed. It is, after all, not the absolute bitumen content which is the indicator of whether the oil-generative window has been reached; the indicator is rather an *increase* in bitumen content. This point is illustrated in Figure 4.4, where the quantity of bitumen at peak generation depends on the kerogen type. If migrational effects are superimposed on generation curves, it may be that no interpretable pattern emerges at all.

A general understanding of the general shape of the curves represented in Figure 4.2 and 4.3 is essential to a meaningful interpretation of oil source-rock data. The exact data used in constructing such a curve will depend upon the particular laboratory analyses carried out. This approach is discussed more thoroughly in Chapters 6 and 7.

3. Bitumen Composition

Consider now the changes in bitumen composition which occur during catagenesis. As has been noted above, the bitumen content of a rock is a function of the degree of catagenesis of the organic material in that rock. The composition of the bitumen should also be a function of the degree of catagenesis, because different types of bitumen molecules are produced from the kerogen during different stages of catagenesis. In general, diagenetic bitumens contain more heterocompounds than do catagenetic ones. The main reason for this is that during early catagenesis, heteroatoms are easily lost from both kerogen and bitumen in the form of small molecules such as CO_2, NH_3, N_2, H_2O, and H_2S. Thus a diagenetic bitumen, or a bitumen formed during the earliest stages of catagenesis, will be of higher average molecular weight than a late-catagenetic bitumen, and its molecules will be more polar.

The elemental compositions of bitumens have not been measured routinely in the West, so no obvious applications to exploration have emerged. Russian geochemists, however, have carried out extensive studies of the compositions of elemental bitumens (Neruchev et al., 1972A, 1978; Pryakhina, 1973; Bazhenova and Gorshkov, 1973). Their conclusions are that the original source material and diagenesis are more important than is catagenesis in determining bitumen composition. It may be concluded, therefore, that there is little likelihood that elemental bitumen analyses will be useful in the evaluation of catagenetic processes.

If bitumens are classified on the basis of the different classes of compounds they contain, parameters are obtained which can be used for studying catagenesis. The four fractions mentioned in Chapter 3, *viz.,* saturated hydrocarbons, aromatic hydrocarbons, resins, and asphaltenes, are present in different proportions in different bitumens. Some of these differences are the result of differing source materials and diagenesis, and some are related to catagenesis. In particular, it has been found that the relative concentrations of saturated and aromatic hydrocarbons in bitumen increase with the thermal maturity of the bitumen (Figure 4.5). This trend is a logical result of both the loss of heteroatoms from bitumen molecules and the increased cracking of carbon-carbon bonds in kerogen during the main phase of oil generation. (Note that it is necessary to consider the complete hydrocarbon profile of a bitumen when making such correlations, because it is an increase in the concentration of hydrocarbons in the bitumen that is being sought. The absolute amount of hydrocarbons in a bitumen is strongly related to source and diagenesis characteristics, rather than solely to catagenesis.)

Bitumen and hydrocarbon profiles have been used for many years by the petroleum industry in an effort to identify oil-generative zones in sedimentary basins. In some cases, such as that shown in Figure 4.5, the bitumen (or hydrocarbon) content increased in accordance with theory when the oil-generative zone was reached, and as a consequence a reasonable and coherent picture of source-rock maturity has emerged (Shibaoka et al., 1973; Vandenbroucke et al., 1976; Albrecht et al., 1976; Albrecht and Ourisson, 1969). In other cases, the bitumen content was strongly affected by such countervailing forces as migration and a change in the type of organic source material, which obliterated the nice picture predicted by Figures 4.3 and 4.4. Furthermore, contamination with drilling fluids can occur readily during drilling and sample recovery, leading to spurious bitumen values. Bitumen and hydrocarbon profiles, therefore, are useful when they work out well, but often they are simply not reliable or informative.

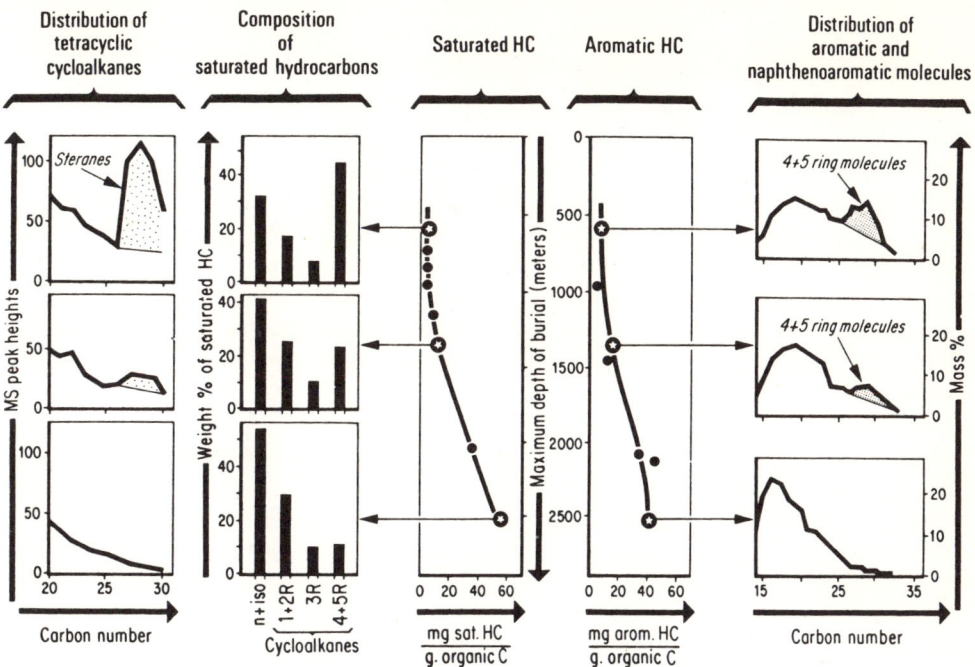

Figure 4.5. Changes in bitumen composition as a function of burial depth in the Paris Basin. The concentration of polycyclic hydrocarbons decreases, while those of saturated hydrocarbons (HC) and aromatic hydrocarbons increase. (From Tissot and Welte, 1978; republished with permission of Springer-Verlag)

Certain classes of compounds present in bitumens have been, or may be, used as indicators of catagenesis. Porphyrins have been suggested as possible indicators of thermal maturity in bitumen (Casagrande and Hodgson, 1974), but their concentrations have not been employed routinely in this role in source-rock analyses. They are used, however, in oil-oil and oil-source rock correlations (see Chapter 7).

The n-alkanes, in contrast, have long been used as indicators of thermal maturity. It was noted two decades ago that while bitumens often preferentially contain the n-alkanes whose molecules have odd numbers of carbon atoms, crude oils usually have equal or nearly equal quantities of the odd- and even-carbon compounds.

Bray and Evans (1961) developed a convenient parameter for measuring the degree of odd- or even-carbon preference in a sample. They defined the *Carbon Preference Index* (*CPI*) as the sum of the concentrations of the odd n-paraffins divided by the sum of the concentrations of the even n-paraffins, with the sum taken over a certain range of carbon numbers. A CPI greater than 1.0 therefore indicates that a bitumen contains more odd-numbered than even-numbered alkanes. The C_{23}–C_{31} and C_{15}–C_{17} ranges have often been used in this capacity.

Figure 4.6 shows n-paraffin distributions for typical immature and mature bitumens and for a crude oil. Figure 4.7 is a compilation of CPI values for numerous samples of each kind. It is clear from these examples that the CPI value gradually approaches 1.0 as catagenesis proceeds. A bitumen that has a high CPI value (greater than 1.2) is therefore thermally immature. CPI values can be used as a negative criterion of maturity, but the converse is not true: some immature bitumens have low CPI's.

Much evidence recently has accumulated indicating that the interpretation of CPI values is not as simple as it once appeared to be. Numerous examples of bitumens and oils having CPI values less than 1.0 (that is, having more even-carbon n-alkanes than odd-carbon ones) have been cited (Welte and Waples, 1973; Dembicki et al., 1976). These distributions cannot have been inherited from the precursor plants, because no organisms are known to produce more even-carbon n-alkanes than odd ones.

Many other immature sediments have been identified in which the CPI is very close to 1.0. Because in immature sediments no thermal catagenesis has yet occurred, a CPI near 1.0 must somehow stem from either the source organisms or some low-temperature diagenetic process.

A paper published in 1977 by a group at the French Petroleum Institute proposed a unified theory that explains the many types of n-paraffin distributions found in bitumens and petroleums in terms of organic source material and diagenetic and catagenetic effects (Tissot et al., 1977). They agree that the CPI of any sample will approach 1.0 as the degree of catagenesis increases. They argue, however, that the initial distributions of n-alkanes, which depend upon the type of organic material from which they were derived, and upon bacterially-induced diagenetic changes, can vary over such a wide range as to make the indiscriminate use of CPI values alone as indicators of maturity very dangerous. The problem of the source and diagenetic influences on CPI values is dealt with in Chapter 7.

As was mentioned in Chapter 3, the average molecular weight of n-alkanes present in bitumens decreases during catagenesis. The position of the maximum in the n-alkane distribution curve (Figure 4.6) should therefore be indicative of the degree of catagenesis. Unfortunately, however, this generalization suffers from the same disadvantage as do CPI values: the position of the maximum in the distribution curve also depends upon the nature of the source organic material.

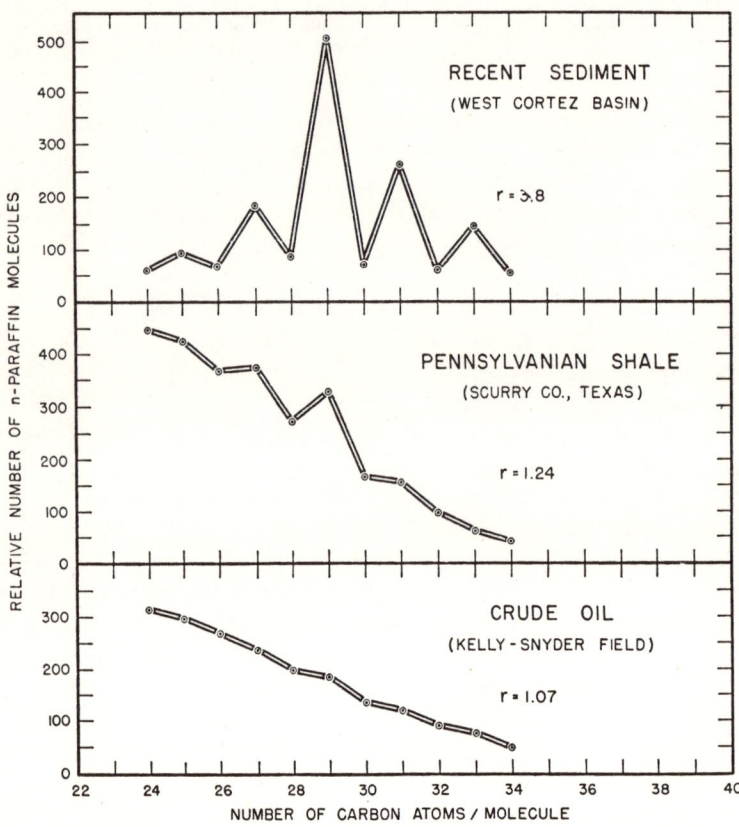

Figure 4.6. n-Paraffin distributions for a Recent sediment, an ancient marine shale, and a crude oil. (From Bray and Evans, 1961; republished with permission of Pergamon Press Ltd.)

Finally, the concentrations in bitumen of saturated hydrocarbons in general, and of *n*-alkanes in particular, increase with increasing catagenesis. Thus *n*-alkanes might constitute 5% of a diagenetic bitumen, 10% of a catagenetic bitumen, and 15% of a reservoired petroleum.

Although in recent years the *n*-paraffins have declined in importance as indicators of thermal maturity, other more complex hydrocarbon molecules have been employed in their stead. The French Petroleum Institute showed a few years ago that polycyclic hydrocarbons undergo some predictable catagenetic changes. Saturated polycyclic compounds appear to undergo gradual transformations that yield acyclic (noncyclic) hydrocarbons and asphaltenes. Thus there is a decrease in the abundance of such compounds as steranes and triterpanes as catagenesis increases. This change is illustrated for three Paris Basin samples in Figure 4.5.

The French workers also found that a gradual conversion of saturated polycyclics to aromatic polycyclics occurs as catagenesis proceeds. Aromatization apparently occurs one ring at a time, so a complex mixture of partially-aromatized molecules is created.

A clever way to show both these changes is illustrated in Figure 4.8. In the example cited, the gradual formation of acyclic compounds is graphically represented, but no increase in aromatic polycyclics is apparent. This approach has not been used routinely in source-rock analyses because the analyses are not simple to carry out. A possible variation would be to look only at certain polycyclic compounds which have been found to be good diagnostic indicators of maturity (Gulyaeva et al., 1978). Alexander et al. (1979) have proposed that the aromaticity of bitumens, as measured by nuclear magnetic resonance (nmr) spectroscopy, is a useful and valid indicator of thermal maturity.

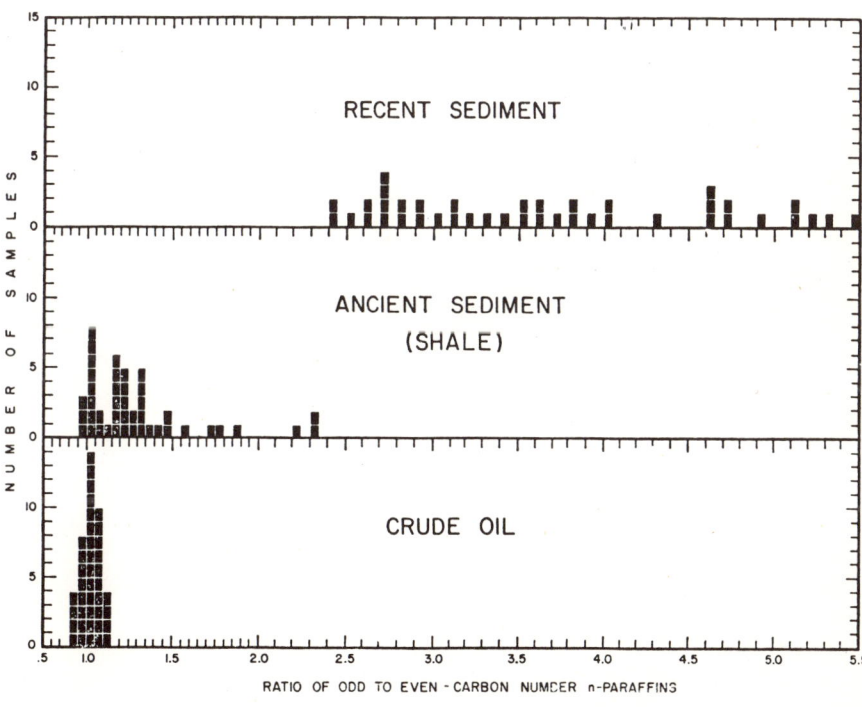

Figure 4.7. *n*-Paraffin distributions in crude oil and sediment extracts. (From Bray and Evans, 1961; republished with permission of Pergamon Press Ltd.)

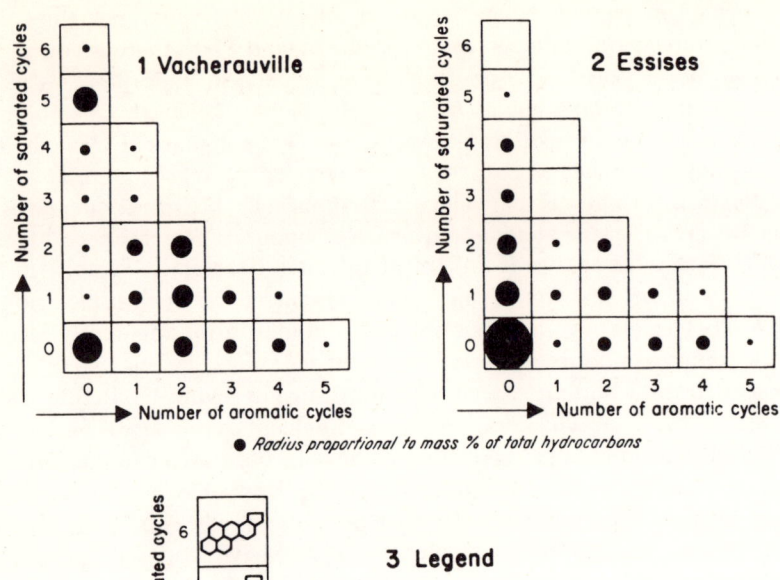

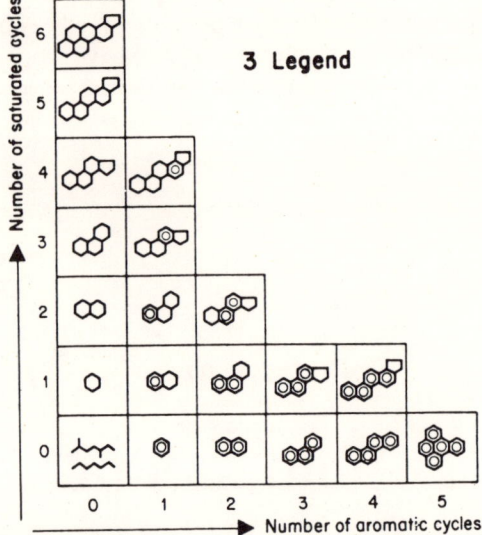

Figure 4.8. Relative abundances of acyclic, naphthenic, naphenoaromatic, and aromatic hydrocarbons in the Lower Toarcian Shales, Paris Basin. The radius of each circle is proportional to the quantity of compounds of each molecular type. Examples of each structural type are shown in the legend. (From Tissot et al., 1971; republished with permission of the American Association of Petroleum Geologists)

The precursors of such complex molecules as steranes and triterpanes are biogenic steroids and triterpenoids (Figure 1.1). Like all biogenic molecules, steroids and triterpenoids are synthesized in very specific ways, and thus have unique structures. An illustrative example is shown in Figure 4.9.

Hydroxyhopane, a triterpenoid of biogenic origin whose structure is shown in Figure 4.9, has its five rings joined in a particular way. At each ring juncture the two angular (noncyclic) substituents stick out on opposite sides of the plane of the molecule. They are thus said to be *trans* to each other. Every hydroxyhopane molecule will have exactly the same geometry as the molecule shown in Figure 4.9.

In a geologic environment, however, chemical transformations can occur which result in subtle but significant changes in molecules like hydroxyhopane. These reactions are thermally-

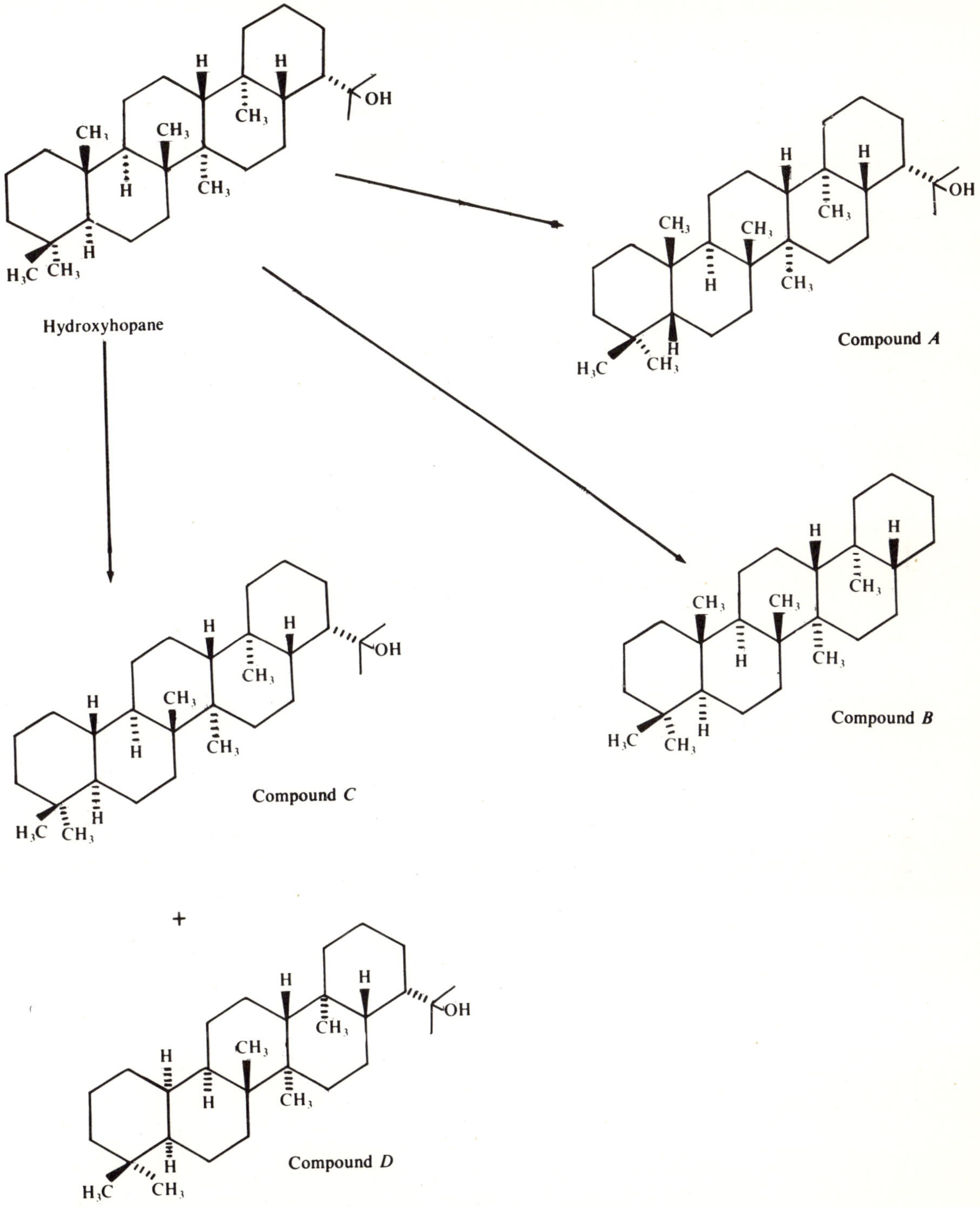

Figure 4.9. Some possible geochemical transformations of hydroxyhopane, a triterpenoid. ▮, substituent points above the plane of the paper; ⚌, substituent points below the plane of the paper.

induced, and they are therefore directly related to catagenetic processes. Among the reactions which might occur are the three shown in Figure 4.9.

Firstly, a hydride ion (H^-) can be removed from one of the ring junctures and later returned to the molecule from the opposite side, yielding compound A in Figure 4.9. Compound A has the two angular substituents at the left-hand ring juncture on the same side of the plane of the ring (*cis* to each other). Compound A is chemically distinct and distinguishable from hydroxyhopane. Its presence is proof that some catagenesis has occurred.

Secondly, substituents like the hydroxyisopropyl group in the right-hand ring can be lost, yielding compound B (Figure 4.9). Finally, an angular methyl (CH_3) group can be replaced by hydrogen. The hydrogen can enter either *cis* or *trans* to the other angular substituent, yielding compounds C and D, respectively.

All of these kinds of reactions can occur at several sites in the molecule. More than one transformation can occur in the same molecule, yielding an exceedingly complex mixture of compounds which differ only slightly from each other. Seifert and Moldowan (1979) have begun to apply combined gas chromatography–mass spectrometry in an effort to analyze these mixtures and extract information on catagenesis from them. The technique holds great promise as a highly sensitive indicator of catagenesis, but routine applications will not be available in the near future because of the complexity and expense involved in such analyses.

In summary, at the present time there are no classes of compounds which can be used as unequivocal measures of thermal maturity in bitumens. High CPI values are indicative of thermal immaturity, but values close to 1.0 may depend at least as much upon source organic material and diagenesis as upon catagenesis. Other parameters have never been developed sufficiently to have gained widespread acceptance, and probably will not be in the foreseeable future. The emphasis of bitumen analyses has shifted away from answering questions of maturity and now is beginning to focus more on solving correlation problems (see Chapter 7).

4. Gaseous and Gasoline-Range Hydrocarbons

Only very small amounts of C_1–C_7 hydrocarbons are generated during the early stages of catagenesis, because formation of these compounds requires the cleavage of high-energy carbon–carbon bonds. During the late stages of catagenesis, and during metagenesis, however, copious amounts of these light hydrocarbons can be produced from both kerogen and bitumen precursors (Philippi, 1975). Numerous attempts have been made to utilize the compositions of gaseous and liquid hydrocarbon fractions as indicators of the catagenetic history of the source material (Philippi, 1975; Hunt, 1979; Evans and Staplin, 1971; Connan and Cassou, 1980). These hydrocarbons are not considered to be part of the bitumen fraction, since because of their volatility they are normally lost during bitumen recovery. Analysis for light hydrocarbons requires special precautions that are discussed in Chapter 6.

As maturity increases, the ratio of acyclic to cyclic hydrocarbons in the light hydrocarbon fraction increases in the same way as it does in the bitumen fraction (Figures 4.8 and 4.5). A variety of different ratios have been used to follow maturation; see, for example, Philippi (1975) and Hunt (1979).

5. Kerogen Structure and Composition

Studies of kerogen composition have long been carried out, with one early objective being the elucidation of a chemical structure for kerogen. This is a very difficult task, and of little practical value on a routine basis, because each kerogen sample is chemically unique. Figure 4.10 shows models for the same kerogen at different levels of thermal maturity. This is the type of structural information that could be attained by laborious analysis of kerogen samples.

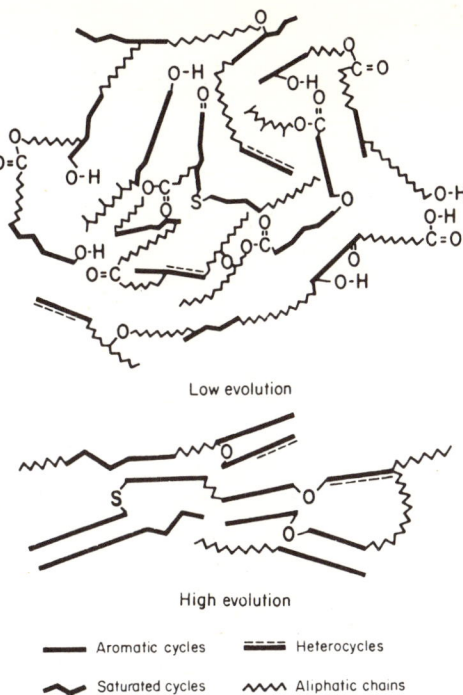

Figure 4.10. Structural model of a Type-II kerogen showing the cross-linking, major component parts, and the differences in packing of organic structures as a function of degree of thermal maturity. (From Tissot and Espitalié, 1975; republished with permission of Editions Technip)

In the last few years, however, it has been realized that such detailed structural information is not required before one can make successful applications of compositional data to petroleum exploration. White (1913) showed long ago that coal composition changes in predictable ways as catagenesis proceeds, and it has since been well documented that the same is true for kerogens (see Chapter 3).

Early diagenesis results in a loss of heteroatoms, mainly in the form of small inorganic molecules such as NH_3, CO_2, and H_2O. Loss of these small molecules produces reactive sites within the residual kerogen molecule, leading eventually to cyclization and aromatization reactions within the kerogen.

As catagenesis proceeds through the oil-generation stage, the molecules being cleaved off from the kerogen matrix now become bitumen-size organic molecules, but the effects on the residual kerogen are very much the same as before: the kerogen becomes even more cyclized and aromatized.

In the late stages of catagenesis, and during metagenesis, cleavage of methyl groups from the kerogen matrix is about the only possible reaction remaining. As each methyl cleavage occurs, the bond, which contains two electrons, is broken in such a way that one electron goes with each fragment. Each of these electrons is therefore unpaired, and any species containing such an unpaired electron is termed a *free radical*. Most carbon free radicals are very reactive and exist for only a short time before they react by pairing up with another electron. In some kerogen molecules, however, free radicals may be stabilized by delocalization over an extended aromatic system.

The process of forming large sheets of fused aromatic rings, such as the structure below, is called graphitization. Graphite is pure carbon in the form of flat aromatic sheets. The flat sheets tend to lie nicely on top of each other; it is this characteristic which allows them to slip over each other and makes graphite a good lubricant. The incipient layering is apparent in the

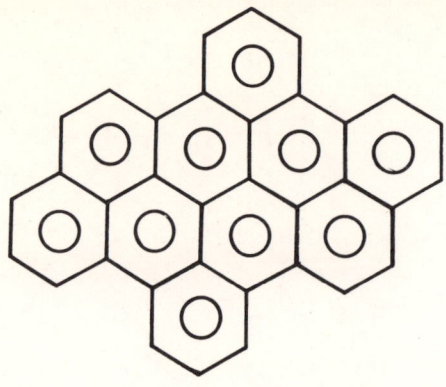

"high maturity" kerogen shown in Figure 4.10. Pure graphite is seldom found in nature, and is probably the result of true metamorphic conditions. Nevertheless, it is useful to draw the analogy between kerogen metagenesis and graphite formation (Powell et al., 1975).

As kerogen becomes more cyclic and aromatic, hydrogen becomes depleted relative to carbon, leading to a decrease in the atomic H/C ratio. This fact is illustrated in Table 4.1, which contains the H/C ratios for a series of low-molecular-weight organic compounds. The H/C ratio therefore contains some fundamental information about kerogen structure, and as a consequence it is useful in estimating the catagenetic status of kerogens. As early as 1913, White had used fixed carbon percentages, which are very similar to H/C ratios, as indicators of maturity in coals. The application of H/C ratios to kerogen maturation and oil generation is discussed in Chapter 7.

Table 4.1. Atomic H/C Ratios of Some Hydrocarbons

Compound	Formula	Atomic H/C Ratio
CH_4	CH_4	4.0
(n-decane)	$C_{10}H_{22}$	2.2
(decalin mono)	$C_{10}H_{20}$	2.0
(decalin)	$C_{10}H_{18}$	1.8
(tetralin)	$C_{10}H_{12}$	1.2
(naphthalene)	$C_{10}H_8$	0.8

In 1969, Staplin published a landmark paper in which he showed that the color of microscopic kerogen fragments was indicative of their degree of catagenesis. During diagenesis, alginites, liptinite, and vitrinite are intially green or yellow, representing the color of the living organism. During catagenesis the color becomes progressively darker, passing through stages of gold, orange, brown, and finally black. Because of their extensive oxidation, inertinites generally are quite dark regardless of the stage of thermal maturation of the sediments. As the macerals become dark brown and black, it becomes increasingly difficult to distinguish the kerogen type accurately. Very few kerogens consist of a single maceral type. Most are complex mixtures with many or all maceral types at least nominally represented.

Because Staplin's color determinations were based on spores or pollen grains whenever possible, this phenomenon became known as *spore darkening,* or the *Thermal Alteration Index (TAI).* Many laboratories immediately began to carry out visual analyses as a means of determining the maturation status of sedimentary sections. Difficulties soon began to manifest themselves, however. The technique is subjective, so constant quality checks must be made. For this reason, several other methods have been developed as substitutes for TAI.

One of these is *vitrinite reflectance,* a much older technique which only recently began to be applied to kerogens. As an important coal maceral, vitrinite had been used for decades by coal petrologists as an indicator of *coal rank* (thermal maturity). It finally was realized that because vitrinite in coal is the same material as vitrinite in kerogen, vitrinite reflectance could be used in the same way for kerogens. The technique involves shining light on a vitrinite particle and determining the percentage of the light which is reflected. The greater the degree of thermal maturity, the greater the reflectance. Vitrinite reflectance values were also used to calibrate H/C ratios, and cross-correlations between relectance and TAI have been successful.

In summary, H/C ratios decrease, TAI values increase (colors change from yellow through brown to black), and vitrinite reflectance values increase as maturity increases. Spore darkening is probably indicative of increased polymerization and aromatization, whereas increased reflectance is related to the rearrangement of aromatic sheetlets into more orderly arrangements.

All three techniques therefore measure important chemical and physical changes in kerogen composition and structure, and their use as maturation indicators is scientifically sound. They have been calibrated empirically for use in defining the levels of kerogen maturity required for oil generation. Conceptually simple, they are well-suited for routine use, and properly employed, they tell a great deal about the past, present, and future of a kerogen. Chapter 6 includes a further comparison and contrast of the three techniques, as does a recent review by Héroux et al. (1979).

Two other techniques have also achieved some measure of acceptance as indicators of kerogen maturity; these are *electron spin resonance* (esr) and *UV-visible fluorescence.* Esr measures the concentration of unpaired electrons in the kerogen. Thermally mature kerogens tend to have a higher concentration of unpaired electrons than do immature kerogens for two reasons. Firstly, the aromatic network in mature kerogens is larger because of increased catagenesis. Secondly, more catagenesis means more cracking, resulting in a greater production of free radicals in the kerogen molecule.

In 1973, Pusey published a paper proposing that the concentration of free radicals in kerogen could be used as an indicator of thermal maturity. He employed electron spin resonance as a measure of the free-radical concentration. Initial enthusiasm for the technique has been tempered by the realization that the free-radical concentration is susceptible to many other poorly understood influences, and may not always correlate well with the degree of catagenesis (Ho, 1979). Applications of esr to kerogen analysis are not common at the present time.

UV-visible fluorescence is a relatively new tool which shows promise as a supplement to vitrinite reflectance for two reasons: it is particularly useful in the range of maturity corresponding to the onset of oil generation, where values of vitrinite reflectance are difficult to measure, and it works best for the liptinite kerogen macerals. Fluorescence is valuable in the analysis of vitrinite-poor samples, such as material of algal origin.

In fluorescence experiments, the sample is irradiated with blue or ultraviolet light, some of which is absorbed by the kerogen molecules. The absorbed energy excites electrons in these molecules to unstable higher energy levels. In time these unstable electronic states decay back to the ground state by reradiating the excess energy as electromagnetic radiation (light). Because of internal-conversion processes, the wavelength of the emitted light is different than the wavelengths that were absorbed. The exact wavelength and intensity of the emitted radiation are measured. It has been found empirically that immature samples have the strongest fluorescence, and that fluorescence intensity decreases with increasing maturity. Furthermore, the wavelength of fluorescence becomes longer (the light becomes redder) with increasing maturity.

Fluorescence is just now being explored as a routine techinque in source-rock analyses. There is much interest in it at the present time, and the reader is encouraged to investigate this topic further in order to decide whether it meets his or her needs. Recent publications (Spackman et al., 1976; Robert, 1979; Ting, 1975; Teichmüller and Wolf, 1977) give an introduction to the principles and applications.

All of the above-mentioned techniques for measuring kerogen maturity are direct indicators of thermal transformations that occur within the kerogen structure. They are, however, only indirect indicators of bitumen formation, the process with which explorationists are most concerned. All of these techniques therefore have the important inherent limitation that one must make inferences about bitumen formation from the behavior of kerogen. It would therefore be advantageous to measure the bitumen-generative capacity of kerogens directly.

This can be done directly by using a technique called *pyrolysis*. Pyrolysis consists of simply heating a sample, in this case kerogen, in the absence of oxygen until thermal decomposition takes place. It mimics natural kerogen catagenesis, and is therefore the most direct indicator available of the bitumen-generative history and potential of a rock.

Pyrolysis has for a number of years been applied occasionally in source-rock analysis. Many early pyrolysis setups were homemade, and little standardization of the technique occurred until about 1977, when the Rock-Eval (discussed in Chapter 6) made its debut on the market. Since then, interest in and application of pyrolysis have increased greatly.

All of the techniques for analyzing kerogens mentioned here have their shortcomings, not only in a general theoretical sense, but also in specific cases. For example, TAI is highly subjective and dependent on the presence of certain types of organic remains, H/C ratios are dependent upon maceral type as well as on maturity (see Chapter 7), and vitrinite reflectance requires the presence of vitrinite. There is ample evidence to suggest vitrinite composition and optical properties are a function of depositional conditions, and Ting (1979) has even questioned the appropriateness of vitrinite reflectance as a measure of oil generation. Pyrolysis conditions in the laboratory are not identical to those that obtain in sedimentary rocks, and there is no guarantee that the same chemical reactions occur in both cases.

Kerogen analyses, like those for bitumen, are therefore not perfect answers to source-rock problems. If possible, more than one technique should be employed. In the first place, the results of the different analytical methods can be checked against each other, and any discrepancies or errors more easily identified. Secondly, some of these analytical procedures actually divulge additional information about kerogen which may be useful in the overall interpretation. For example, reflectance work can yield information about the proportion of reworked (second-cycle) material in the vitrinite. Visual analyses carried out in conjunction with TAI measurements not only help to assess the contribution of reworked kerogen, but also give important information on the organic macerals present. Rock-Eval pyrolysis also yields information about the type of organic matter present (see Chapter 6). Properly interpreted and used in conjunction with other techniques, kerogen analyses are an invaluable part of modern source-rock evaluations.

5 MIGRATION OF OIL AND GAS

1. Introduction

Theories of migration have been plentiful through the years, each coming into fashion for a while and then dropping from favor. A few have even had resurgences of popularity, but despite the interest in the process of migration, and the obvious economic importance of understanding it, we are only beginning to comprehend how migration actually occurs.

One reason that an understanding of migration has been such an elusive goal is that the mechanistic aspects of the phenomenon have too often been oversimplified. For years, geochemists and petroleum geologists have sought a single migrational mechanism which could explain all oil and gas migration. Numerous theories—micells, continuous phase, droplets, solution, clay diagenesis, carbonate diagenesis, etc.—have been suggested, but all are deficient in some respects. To evaluate these and other possible mechanisms, the phenomenon of migration must first be clearly defined. The various mechanisms that have been proposed can then be discussed on the basis of the factors that might influence migrational efficiency. Finally, it must be kept in mind that several different migration mechanisms may be operative under different geologic conditions.

Primary migration is the movement of bitumen out of the fine-grained source rock and into a more permeable conduit. *Secondary migration* is the movement of bitumen (or petroleum) through the permeable conduit into the reservoir. *Accumulation* is the termination of secondary migration at the reservoir.

There are no definite distance constraints on primary migration. It may be a very short-distance process in cases where there is a favorable source rock–reservoir relationship. Examples of such cases are a fractured shale reservoir like the Miocene-age Monterey Formation in California, where the source rock is also the reservoir, and thin beds of interfingering sands and shales, such as those commonly found in deltaic sequences. In other cases, particularly where the fine-grained rock consists of a single massive formation, the bitumen generated in the center of the formation will have a long primary migrational pathway to traverse.

In general, primary migration is limited to a few tens or perhaps hundreds of meters, while secondary migration undoubtedly can occur over much longer distances. The Athabasca tar sands in Canada, the Eastern Venezuela heavy oils, and some of the supergiant oil fields in the Middle East require drainages over vast areas, with lateral migrations of up to several hundred miles. These cases are exceptional, because they represent some of the largest accumulations in the world, but they show that given the necessary lateral continuity of migration conduits, secondary migration will continue until a trap is reached, or until the organic material is destroyed by oxidation. Claypool et al. (1978) have considered the migration distances required if Paleozoic oils from central Wyoming were generated in the Phosphoria Formation. Distances range from 60 to 250 miles, depending upon the assumed migrational efficiency.

In more typical situations the extent of lateral migration is probably a few miles. In the Los Angeles Basin, for example, the main depocenter is located 10 to 20 miles from most of the largest fields. Proximity of traps to the sites of oil generation obviates the need for extremely long migration.

Almost all lateral migrations will also have a vertical component, because subsurface fluids move toward regions of lower potential, which usually means upward, and because the migration conduits themselves lead upward. If the conduit is a porous layer, it normally will approach the surface as it progresses from the center of a basin to its flanks. Migration will in general be parallel to the bedding, except where active faults provide pathways for the oil to cut across strata.

The limits to the extent of vertical migration are determined by the geology of each province. If no active faults existed during the period of migration, then the dip of the bedding determines the rate of vertical movement. If active faults were present, however, vertical migration may be much more dramatic in scale.

The reason that a distinction is made between primary and secondary migration is that the fluid dynamics are very different in the two cases because of different degrees of interaction of interstitial fluids with mineral surfaces. Fine-grained rocks which are the hosts for primary migration maximize fluid–mineral interactions, while coarse-grained rocks and fault surfaces, through which secondary migration occurs, minimize these interactions. Secondary migration also is likely to be much more rapid than primary migration.

2. Mechanisms of Primary Migration

2.1. IN AQUEOUS MEDIA

Three distinct mechanisms by which organic molecules could migrate though water-wet rocks have been proposed. The distinguishing features of each are the size of the organic units which exist within the pore fluid. These three mechanisms can be called true solution, micellar or colloidal migration, and migration as droplets or globules. The approximate size ranges of each of these organic units are given in Table 5.1. The behavior of each of these kinds of hydrophobic organic units in a water-wet system is distinct, and worthy of a detailed analysis.

True solution occurs when the organic molecules are separated from each other and are independently solvated by water molecules, as is shown in Figure 5.1. Migration in true

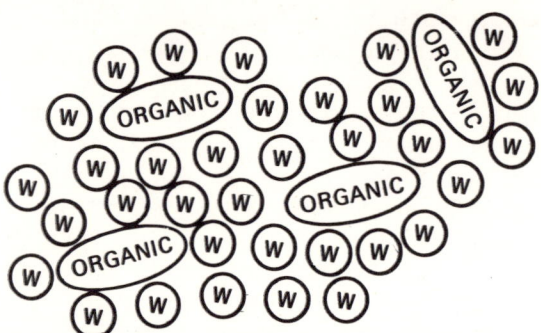

Figure 5.1. True solution of organic molecules in water.

solution had been largely discredited by 1970, in part because McAuliffe (1966) had shown that at 25 °C the water solubility of petroleum hydrocarbons, especially the heavier ones, was far too low to account for transport of the vast quantities of petroleum found in reservoirs today. In 1973, however, Price presented new data which showed that hydrocarbon solubility increases exponentially with temperature, and that the solubility of

Table 5.1 Approximate Size Ranges of Organic Units during Migration in Aqueous Media

Unit type	Unit diameter range (Å)
true solution	4–10
micells (colloids)	10–10,000
droplets	> 10,000

hydrocarbons at 100 °C is many times greater than it is at 25 °C. Furthermore, the heavy hydrocarbons show the greatest relative solubility increase with increasing temperature. The net effect of Price's work was to rekindle interest in true solution as the dominant mechanism in primary migration.

Price himself, however, changed his ideas about the role of true solution after calculating the volumes of water necessary to accomodate all known accumulations of hydrocarbons in reservoirs. He concluded that temperatures of 100 °C were insufficient to solubilize the required amounts of bitumen, and that therefore migration must occur at temperatures of about 200 °C (1976). This has led Price subsequently to pursue his theory of the "hot and deep" origin of petroleum, an idea with which most petroleum geochemists strongly disagree.

There are other problems with true solution besides volumetric considerations. The distribution of hydrocarbons in petroleum is very different from that which one would expect on the basis of water solubility alone. Light aromatic hydrocarbons such as benzene and toluene are more soluble than saturated hydrocarbons by factors of 100 or so, yet these compounds are depleted in petroleum relative to bitumen. Similarly, many heterocompounds are much more soluble in water than are hydrocarbons, but they are usually only very minor constituents of petroleum.

Saturated hydrocarbons, on the other hand, undergo a significant enrichment during migration, yet these compounds are among the least water-soluble components of petroleum. The composition of petroleum seems therefore to be approximately inversely related to the solubility in water of the individual components. This fact may be of significance in understanding the process of accumulation, but it does not in any way support true solution as a migration mechanism. A tentative conclusion might therefore be that true solution may be an important mechanism for some of the lightest, most soluble hydrocarbons, but that it cannot be a major contributor for heavy components.

It has been suggested that hydrophobic compounds form small aggregates, or *micells*, in aqueous media to minimize the unfavorable hydrophobic interactions. Fatty acids can do this particularly easily, because one end (the carboxyl group) is hydrophilic and the other end (the alkyl chain) is hydrophobic. Such a micell is shown schematically in Figure 5.2. The hydrophilic ends are in contact with the aqueous medium, while the hydrophobic portions lie within the micell, and interact only with each other. Mixed micells also exist in which a variety of surfactant molecules such as phenols and carboxylic acids protect hydrocarbons. These micells are the size of colloidal particles.

Baker (1959) and Cordell (1972, 1973) have proposed migrational mechanisms that involve micells. The theory has many attractive features, because oils contain significant quantities of

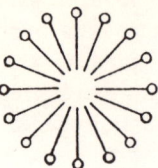

Figure 5.2 Micell composed of fatty acid molecules. Round ends represent carboxyl groups (hydrophilic); lines are alkyl groups (hydrophobic).

surfactants like carboxylic acids (Seifert and Howells, 1969). Furthermore, micells could increase dramatically the volumes of hydrocarbons transported in water. Finally, solubility data for *n*-alkanes in water suggest that the *n*-alkanes with more than 10 carbon atoms form micells rather than dissolving in true solution. Figure 5.3, which shows that the C_{20+} *n*-alkanes are much more soluble than would be expected on the basis of the behavior of C_1–C_8, supports the micell theory. Interestingly, the distributions of *n*-alkanes in crude oils (Figure 5.4) are quite similar to the solubility curve shown in Figure 5.3.

On the basis of the above evidence it seems plausible that micells could be important in petroleum migration. The problem with the micell theory arises when micell sizes are compared with pore diameters that are available during primary migration. Baker (1962) has suggested that ionic micells large enough to carry hydrocarbon molecules inside them would have average diameters of 60 Å. Neutral micells would have to be much larger, perhaps 5000 Å. Because pore diameters have been reduced to about 50 Å in shales at depths of 2000 meters (a plausible depth for oil generation), primary migration by ionic micells is not feasible at greater depths. Neutral micells apparently could not migrate under any reasonable geologic conditions.

There are two other problems with the micellar theory of migration. In the first place, electrostatic repulsion between negatively-charged clay surfaces and negatively-charged ionic micells would prevent the passage of the micells through narrow pores. Secondly, a very large excess of polar compounds is required to transport a given volume of hydrocarbons within an ionic micell. The fact that such polar compounds are not found in reservoired petroleum mitigates strongly against the micellar theory of primary migration.

Migration as globules or bubbles which are somewhat larger than the largest micells has also been proposed, but this proposal seems to have little to recommend it as a mechanism for

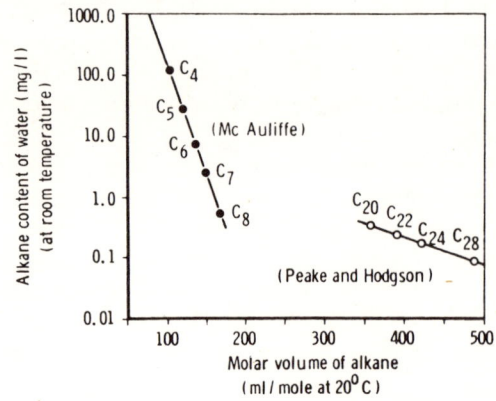

Figure 5.3. Solubility of *n*-alkanes in water. The unexpectedly high solubilities of the heavier members are attributed to micell formation. (From Peake and Hodgson, 1966; republished with permission of American Oil Chemists' Society)

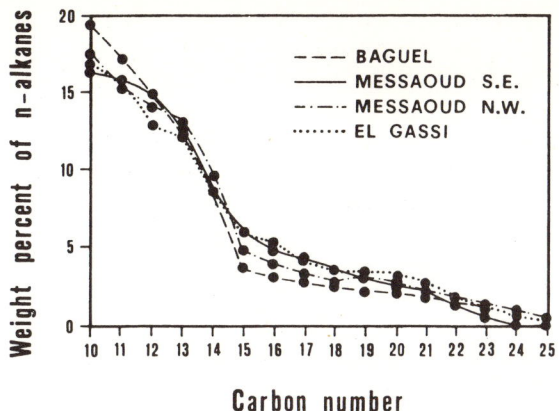

Figure 5.4. *n*-Alkane distributions in four crude oils. (After Price, 1973)

primary migration. Globules will certainly be larger than pore diameters in fine-grained rocks, and they must therefore undergo deformation if they are to pass through the openings (Figure 5.5). Calculations by Hobson (1973) indicate that rather high entry pressures are required to force the globules through pores. Although these pressures could conceivably be reached in rocks, such pressures could also cause extensive microfracturing, and thereby precipitate migration through fractures rather than through pores. Tissot and Welte (1978) therefore conclude that primary migration does not occur via droplets or globules.

The above analysis has not yet considered the relationship between hydrodynamics and the movement of bitumen. *A priori*, three possibilities exist: diffusion of bitumen through a static water medium, movement of bitumen as a result of hydrodynamic transport, and movement by a combination of diffusion and hydrodynamic flow.

Calculations suggest that, given geologic time, diffusion along a concentration gradient could account for the transport of enough bitumen to achieve moderate-sized petroleum deposits, but not supergiant accumulations. Diffusion along pressure gradients and along thermal gradients has also been proposed, but these processes probably are relatively unimportant. Diffusion, therefore, seems to be an auxiliary process in primary migration, rather than a dominant one.

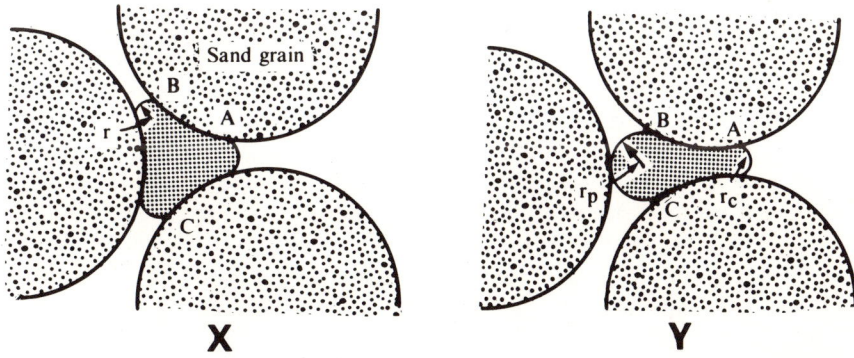

Figure 5.5. Movement of an oil droplet through a small pore, showing deformation required for passage through a pore constriction. (From *Geology of Petroleum,* Second Edition, by A. I. Levorsen. W. H. Freeman and Company. Copyright © 1967.)

Price (1976) has shown that immense volumes of water are required if true solution is important in primary migration. Where does all this water come from? It has been proposed that it is obtained by the dehydration of clay, but volumetrically this would appear to be inadequate, and in any case the timing of clay dehydration does not seem to be favorable (Waples, 1980). Ordinary water of compaction has also been suggested, but there probably is not enough of it. Furthermore, as Emery and Rittenberg (1952) and Hobson (1961) showed long ago, and Bonham (1980) recently reemphasized, the net flow of compaction water in a subsiding basin is actually downward with respect to the center of the earth, rather than upward. If waters expelled during compaction carried migrating bitumen, petroleum reservoirs would occur at depths greater than those at which the bitumen was formed.

Migration can occur in a continuous organic phase, and Dickey (1975) and Hill (1959) have suggested that as compaction reduces pore volume, the organic material begins to occupy a very substantial fraction of the total pore space. With a reasonably rich shale and substantial compaction, the organic material may form a continuous network within the shale. Within this continuous network, mobile bitumen molecules would be free to move under the influence of physical stresses. The problems which exclude micellar and globular migration would not arise, nor would the solubility problems which plague the true solution theory.

Philippi (1974) has taken a stand for continuous-phase migration, and evidence is beginning to support this hypothesis. Carbonates, for example, tend to form continuous organic layers as a consequence of recrystallization. Organic-rich shales should have no difficulty in forming organic networks.

Average shales, on the other hand, would have difficulty concentrating their one-percent-content of organic carbon into a continuous phase, even when porosity was reduced to a very low level. For this reason, it seems that continuous-phase migration is likely to be dominant only for rather rich shales (perhaps with organic carbon values of 5% or more), and in some carbonates.

Some evidence supports this view. In Utah, veins of solidified bitumen are plentiful, and have even been mined commercially in the past. Some of the veins attain thicknesses of five meters (Hunt, 1963). These bitumens may be the remnants of continuous-phase migrations of past ages. The source rocks for these bitumens probably lie in the extremely organic-rich Uinta, Green River, and Wasatch Formations. No direct evidence has been found that continuous phase migration has occurred in regions without such rich source rocks.

As the reader will have appreciated by now, getting organic matter through small pore openings is not easy, no matter what the mechanism. It has therefore been proposed that primary migration can also occur through microfractures. The advantage in calling upon microfractures is that at the moment of fracturing the openings available to mobile organic molecules are much larger than pore openings. Relatively large molecules, such as porphyrins and asphaltenes, could thus migrate much more easily than they could through pores.

Snarsky (1962) showed that microfracturing is common in compacted shales and carbonates. Tissot and Pelet (1971) and Lewan et al. (1979) have reproduced this effect experimentally. Migration via a system of microfractures would of course be a discontinuous process, like movement along a fault. It would proceed in a series of pulses, with pauses while fluid pressures built up again. The fluid phase migrating through microfractures could be either aqueous or organic, depending upon the richness of the source rock.

Mills (1923) noted that when gas was dissolved in an oil–water mixture under pressure and the container then punctured, the gas expelled through the hole carried both oil and water with it. The same process might obtain in microfractures, and if it does, it could facilitate the movement of the heavy components of petroleum.

Because microfracturing is one of the newest arrivals on the migration-hypothesis scene, it has not yet had a chance to be thoroughly evaluated. It is, however, an attractive idea that should be examined carefully.

3. Timing of Primary Migration

There are few documented cases of migration involving diagenetic bitumen. Gilsonite veins in Utah and asphaltic bitumen/oil deposits in the Monterey Formation in California are examples of these unusual occurrences. In general, however, migrating organic fluids are catagenetic, and it therefore follows that most migration occurs during or after catagenesis. Tissot and Welte (1978) state that most primary migration occurs at burial depths between 1500 and 3500 meters, which is also approximately the depth range for bitumen generation from kerogen. But how much delay is there between generation and migration? The answer to this complex question lies, I believe, in the particular migrational mechanism operative in each case. Philippi (1965, 1974) has suggested that the first-generated bitumen is adsorbed on kerogen sites, and that migration can occur only when all adsorption sites have been filled. This view suggests that some delay must occur between generation and migration of bitumen. Philippi's constraint would seem to apply equally well to solution- or continuous-phase migration. Migration through microfractures would also require a delay, because microfracturing will not occur until internal fluid pressures (caused at least in part by kerogen catagenesis) have built up sufficiently.

Other mechanisms proposed for migration require special timing. The dehydration of clay has been suggested (Powers, 1967; Burst, 1969) as a source of water to serve as a migrational medium. If this idea is correct, then migration could not occur until the dehydration of clay made the water available. Recent work, however, indicates that the dehydration of clay usually precedes the formation of bitumen, and that even if true solution were an important migration mechanism, it could not be dependent upon clay dehydration (Waples, 1980).

It therefore seems likely that migration is somewhat delayed after bitumen generation has begun. Figure 4.4 supports this view, for if migration commenced at the same time as bitumen

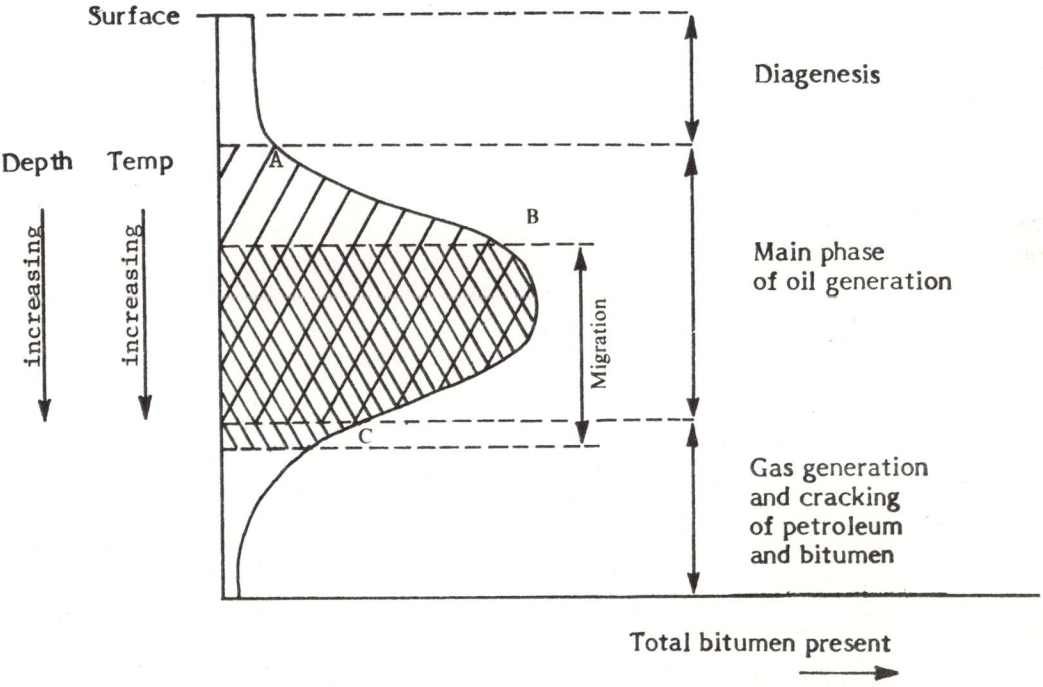

Figure 5.6. Comparison of timing of oil generation and migration.

generation, the bitumen concentration would never reach a maximum, but rather would remain at a low level—bitumen would migrate out of the source rock as soon as it was generated. Once Philippi's adsorption sites are saturated, however, bitumen migrates about as fast as it is generated.

Seen from this point of view, it is apparent that migration will cease shortly after bitumen generation ends. Figure 5.6 shows the correlation between migrational timing and bitumen generation.

It can be seen from Figure 5.6 that the migration window (that is, the depth, temperature, and time limits for migration) overlaps, and in fact is slightly narrower than, the bitumen-generation window. In most cases, the period required for generation and migration will amount to no more than a few million years.

From a practical standpoint, then, the generation and migration of bitumen can be considered to be nearly synchronous. If the depth and time at which bitumen generation occurred can be identified, then the time of its migration can also be determined. This synchrony is very helpful when the time-temperature model for oil generation is applied (see Chapter 8).

It will not always be the case, of course, that generated bitumen is able to migrate out of the source rock. The model therefore not only predicts the time at which migration must occur, but it also predicts the consequences if migration cannot take place.

Momper (1978) has imposed rather specific requirements for migration to be initiated, stating that a minimum of 15 barrels of bitumen per acre-foot of rock (50 million barrels per cubic mile) must be available before any can be expelled from the source rock. If permeability or fluid-dynamic considerations prohibit its movement within a short time after it is generated, the bitumen will crack further to yield gases and gasoline-range hydrocarbons.

4. Mechanisms of Secondary Migration

Older theories about secondary migration assumed that bitumen moved in true solution to the reservoir, where changing conditions of temperature, pressure, and salinity forced it out of solution and allowed the petroleum to be trapped in the reservoir. One weakness of this theory is that it requires that the bitumen come out of solution at precisely the right place to be trapped. Furthermore, it implies that the changes that cause the bitumen to separate from the solution are sudden and drastic. It is now known that changes in temperature and pressure within migrational conduits are gradual, and that although changes in salinity can be abrupt, they alone could not cause huge quantities of oil to separate from solution.

It is much more reasonable to assume that oil globules form at a much earlier stage of secondary migration. There are several possibilities, none of which can be favored on the basis of current knowledge. The first possibility is that globules could form suddenly at the interface between the source rock and the secondary migration conduit in response to the drastic changes in mineral–fluid interactions that must often occur there. A second possibility is that globules could form all along the secondary migration path as gradually changing temperature, pressure, and salinity result in the coalescence of smaller organic units. The third possibility is that globules could already exist as the bitumen emerges from the source rock. This would be the case if primary migration occurred either as a continuous phase or via microfracturing.

In all of these cases, secondary migration will involve mainly droplets or globules of oil. Migration of droplets through a porous conduit takes place much more readily than it does in a source rock. There is space in the larger pores to accomodate relatively large droplets, and the increased pore diameters will lower capillary entry pressures and permit the droplets to move more readily.

There is some recent indirect evidence that either micells or droplets play an important role in secondary migration. Bray and Foster (1980) have shown that the presence of CO_2 and hydrocarbon gases greatly enhances the ease with which crude oil migrates through sandstone. This effect cannot be explained by enhanced solubility, and it therefore must involve formation of some multimolecular organic structures.

Globules move in response to buoyancy. They are less dense than water, and therefore will tend to move upward with respect to the water. Very small globules do not have sufficient buoyancy to rise, however, and they therefore remain immobile with respect to the water until several can coalesce to form a single globule with the necessary buoyancy. The many factors influencing the rate of buoyant rise are discussed by Tissot and Welte (1978). Superimposed on buoyant movement is hydrodynamic flow, which can either add or detract from the net effect of buoyant flow, depending upon the direction of the hydrodynamic gradient. Tissot and Welte (1978) also discuss this point.

5. Accumulation

According to the globule theory of secondary migration, the physical processes involved in the accumulation of oil are quite simple. Because hydrocarbon molecules are moving in response to buoyancy requirements, their movement can be stopped by preventing their upward motion. This is like trapping a helium-filled balloon against the ceiling. When secondary migration is visualized in this way, it is easy to explain why reservoirs are located where they are. Consider, for example, a pinchout (Figure 5.7).

Fluids fill the conduit throughout its length, but no bulk flow occurs because the upper end is sealed. The only movement of fluids is a gradual upward movement by the more buoyant oil globules, which in turn displace water downward.

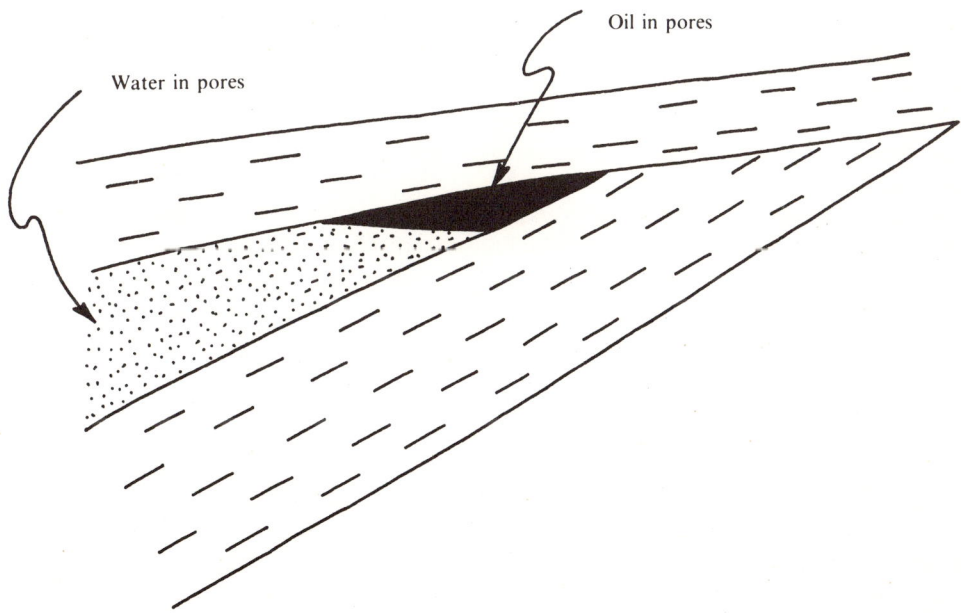

Figure 5.7. Pinchout of a migrational conduit. The upper part is filled with oil and the lower part with water because of buoyancy differences. There is no bulk fluid flow, because the upper end is sealed. Oil globules move upward, displacing water downward.

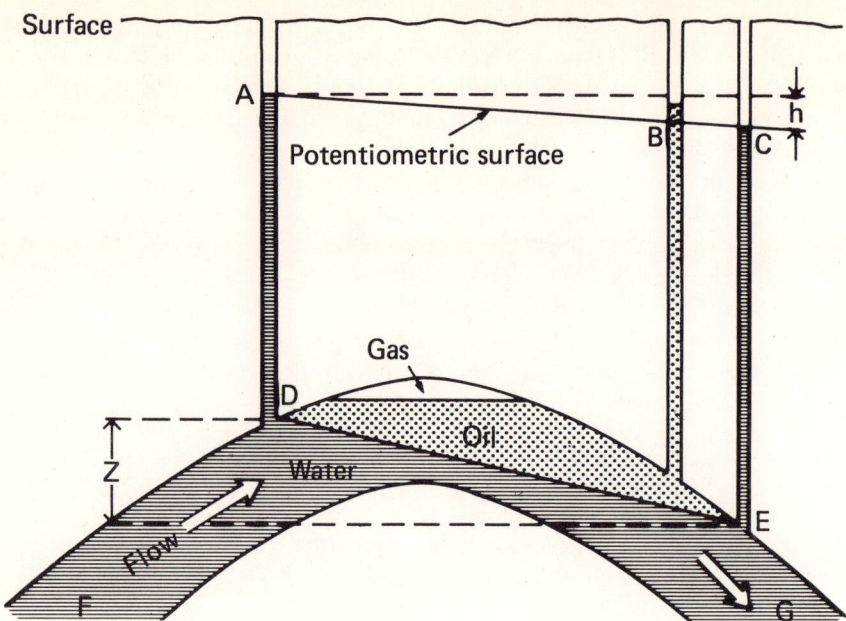

Figure 5.8. Entrapment of buoyant oil and gas at the tops of reservoirs. Hydrodynamic flow creates a slope in the oil-water contact. (From *Geology of Petroleum*, Second Edition, by A. I. Levorsen. W. H. Freeman and Company. Copyright © 1967)

A second example is a normal anticlinal reservoir (Figure 5.8). The more buoyant oil globules will gradually fill the highest points in the reservoir. No dramatic physical or chemical processes are required; accumulation follows smoothly from the laws of physics.

Accumulation is not always the final chapter in the story of oil generation and migration, however. No cap rock forms a perfect seal, nor is any trap exempt from possible destruction by faulting or erosion. Many examples exist of breached reservoirs which once held oil and now contain only tarry residues or heavy oil. Two aspects of this phenomenon are of particular interest to us here.

Firstly, migrating hydrocarbons often form their own reservoir seal. Formation of stratigraphic traps sealed by tar occurs frequently where inspissation from biodegradation and evaporation have occurred, usually at or near the surface. Tar sealing therefore slows down the loss of oil where reservoir seals have been destroyed. Secondly, all reservoirs are actually dynamic systems. A hydrocarbon accumulation forms when the rate of in-migration exceeds the rate of leakage, and can persist after in-migration ceases only if the rate of leakage is low. Rates of leakage are certainly very different in the many existing reservoirs. Areas such as Southern California, which have major active seeps, are quite leaky, while in areas like the Midcontinent, oil generated and trapped in the Paleozoic persists today. Exploring for old oil in a leaky area is likely to be fruitless because of the evanescence of such accumulations.

A graphic example of how mobile hydrocarbon accumulations can be in leaky areas was discussed by McCulloh (1969), who examined the Dos Cuadras Field, site of the Santa Barbara Channel oil spill. The reservoir seal was a thin, porous siltstone which is less than 300 feet thick above the shallowest major reservoir, and which itself contained producible hydrocarbons. The sediment package therefore represented a metastable system which, when disturbed by drilling, discharged hydrocarbons violently. Left undisturbed, these accumulations would probably dissipate in a geologically short time by natural seepage.

6. Accumulation Efficiency

For many years, geologists and geochemists have considered petroleum migration to be a very inefficient process, with perhaps one or two percent of the bitumen generated during catagenesis eventually being trapped in a reservoir as petroleum. This hypothesis was in conflict with the mass-balance requirements for supergiant accumulations like the Eastern Venezuela heavy oils, the Ghawar field in Saudi Arabia, and the Athabasca tar sands. It was therefore assumed that migration must be highly efficient in these unusual cases, but low in normal oil provinces.

Recently, however, mass-balance calculations for relatively well-studied but otherwise unexceptional oil provinces suggest that migrational efficiencies are much higher than was previously believed. The critical factors that determine how efficiently migration occurs in a given area are, predictably, geologic in nature.

First and foremost among these factors is the existence of continuous lateral conduits leading from the center of the basin, where the most intense oil generation will be, to the flanks, where traps are likely to be found. These conduits must be continuous in both time and space. A highly focused conduit, like an ancient turbidite channel, can be highly effective in this regard. If lateral continuity is poor, accumulations are likely to be smaller, and much of the oil may never reach the basin flanks. Instead it is trapped within the basin, where subsequent cooking will convert it to gas.

The relative timing of structure growth and oil generation also must be favorable if entrapment is to occur. If at any time the structure is too small to handle the amounts of oil being poured into it, then oil will spill out and migrate onward, with a correspondingly reduced accumulation efficiency (unless it is trapped later by a different trap). A careful analysis of timing is essential to any regional evaluation; the model discussed in Chapter 8 can be useful in this regard.

A third important factor in migrational efficiency is the presence of active faults, such as growth faults, during the period of oil generation. These faults may discourage any long-distance lateral migrations, but on the positive side, they may lead to numerous smaller accumulations that are not associated with structural traps on the flanks of the basin. Migrational efficiency is probably higher in many such cases, of which the Gulf Coast is an example.

Hybrid situations can also exist in which an early and continuous lateral conduit is subsequently cut off by faults which break its continuity and drain off the oil which migrates through it. Knowledge of the relative times of faulting and generation would be important to an explorationist who was evaluating such a basin.

6 SAMPLE ANALYSIS

1. Introduction

Although most people involved in oil exploration do not directly participate in the organic geochemistry laboratory analyses, it is nevertheless very important that they be familiar with the procedures used. Any analytical method has strong and weak points, and an understanding of possible problem areas in analytical procedures can often help one to deal with apparently anomalous data. A variety of analytical errors can occur, and unless data quality is carefully monitored, the errors will not be noticed.

2. Sample Quality

The problem of data quality actually begins long before the samples arrive in the laboratory. Drilling crews often seem to consider sample requests from organic geochemists to be a nuisance; many times they have no idea why a particular kind of sample is requested. As a consequence, geochemists must be particularly careful to verify that the samples received are actually those requested.

Contamination after sampling can be a significant problem. Plastic bags often have been used as sample containers, but plasticizers often get into the samples. Before plastic containers are used, tests should be run to confirm their suitability. Canned, sealed samples are best, but these precautions are seldom taken. Volatile components will be lost from any samples exposed to the atmosphere. Samples which have been stored in core repositories for extended periods are seldom suitable for bitumen analyses.

Microbial activity which occurs after sampling often can affect the organic material in samples, especially in unlithified sediments or those which are very rich in organic compounds. As an illustration of how pervasive a problem this can be, I have seen Deep Sea Drilling Project sediment samples, which were stored at 4 °C within hours after recovery, develop growths of fungi or mold in a few weeks. None of the samples contained more than 0.2% organic carbon. Microbial transformations will generally be less of a problem for rocks because of the inaccessibility of the organic material.

Samples obtained during drilling are generally either conventional cores, sidewall cores, or cuttings. All of these are susceptible to contamination introduced during sample recovery. The potential for contamination is of course much more severe for cuttings, which consist of small fragments of rock broken off during drilling.

Conventional and sidewall cores and cuttings can all become impregnated with material mixed into drilling fluid. Diesel fuel and other petroleum products are occasionally used. In such situations an accurate log of drilling-fluid additives is invaluable to the petroleum geochemist, but such logs are seldom available. Often there is no official record of the exact nature of such additives, but the amazingly high bitumen contents of the contaminated samples will tolerate no other explanation. Surprisingly, the cores often are deeply impregnated with the contaminants, so the problem cannot always be solved merely by removing the surface layer of rock. Contaminated samples are generally useless for bitumen analyses, but they may still be acceptable for kerogen studies.

Walnut hulls, rubber, and other solid organic debris are also sometimes dumped into wells during drilling, and these substances can complicate kerogen analysis. Careful microscopic examination, however, can usually identify problem samples. The only real solution to the problem of contamination is to avoid it. In this respect, it would be helpful to include drilling engineers in geochemical planning.

One of the most serious contamination problems comes from mixing of different lithologies in the course of transport of the cuttings to the surface. Almost all cuttings samples require laborious hand picking under a microscope if anything approaching a homogeneous sample is desired. Even then, the homogeneous sample may not always represent the intended interval.

3. The Analytical Scheme

Once the samples have been obtained and their authenticity and suitability have been verified, the particular analyses to be carried out depend upon each laboratory's approach to source-rock and oil interpretation and correlation. Time and money constraints often prohibit a complete analysis of all the samples available. In such cases a preliminary screening can indicate which samples are worthy of further, more detailed examination.

A common screening technique is measurement of total organic carbon content. In one method, the crushed rock sample is burned in a LECO furnace, and the evolved CO_2 is measured. The amount of CO_2 formed depends directly upon the total carbon content, both organic and inorganic. Next, another portion of the crushed rock is digested in hydrochloric acid to remove carbonate (Eq. 6.1), and the amount of CO_2 evolved is measured. This quantity is related to the inorganic carbon content. Finally, organic carbon is calculated by subtracting inorganic carbon from total carbon. This method is good except for carbonates. In those cases the organic carbon content is a small number derived by taking the difference between two large numbers (total carbon and carbonate carbon), and thus may contain a large percentage error. Caution should be exercised in interpreting organic carbon values for carbonates unless duplicate analyses have been run.

$$2\ HCl + CaCO_3 \longrightarrow CaCl_2 + H_2O + CO_2 \qquad (6.1)$$

3.1. BITUMEN ANALYSIS

If bitumen analyses are desired, the next step is extraction of bitumen from the rock. After the sample has been crushed to about 100–200 mesh, there are two main kinds of extraction available. The traditional one is solvent extraction, which is usually carried out with a Soxhlet condenser. This allows continuous extraction with fresh solvent over whatever time period is desired (usually 8–24 hours). A variety of solvents has been used. The Russians have used chloroform for many years. Until recently a mixture of benzene and methanol was the most popular in the West, but health considerations are leading to a changeover to chloroform. Ultrasonic extraction was popular five to ten years ago, but it has not been used much since.

One definite advantage that ultrasound has is its speed (20 minutes), but the results obtained do not seem to be any more reproducible than those obtained from Soxhlet extraction.

The next problem is to remove the solvent from the extracted bitumen. This is usually accomplished by low-temperature evaporation, which often is effected by passing a stream of nitrogen gas over the solution. During the evaporation process, the more volatile bitumen molecules evaporate with the solvent, and the remaining bitumen is severely depleted in compounds that have fewer than 15 carbon atoms. This complication is unavoidable, and it means that the bitumen that we subsequently weigh and analyze is not identical to the bitumen actually present in the rock. The bitumen actually isolated and measured gravimetrically is often referred to as the C_{15^+} bitumen.

If one wishes to compare extraction data obtained from different groups of samples, it is essential that standard conditions of solvent type, extraction time, particle size, and so forth be maintained. Even so, results are not always as reproducible as one might desire.

Thermal extraction is a relatively recent development which shows great promise. A crushed rock sample is exposed to high temperatures which mobilize its bitumen and allow it to be swept out of the sample by a gas flow and collected and analyzed. The temperatures used in thermal extraction are lower than those required to decompose kerogen, however. The most important application of thermal extraction is the Rock-Eval instrument developed by the French Petroleum Institute, which will be discussed in detail later in this chapter. Bitumen obtained by thermal extraction in the Rock-Eval is analyzed directly within that instrument, so the remainder of this section applies only to solvent-extracted bitumen.

After the solvent has been stripped from the extracted bitumen, the next step is to separate the different classes of compounds in the bitumen. The asphaltenes are first precipitated by addition of an excess of pentane and filtered out. The remaining bitumen is then separated by column chromatography. A vertical glass tube is filled with a slurry containing either silica gel or alumina, both of which have considerable adsorptive powers. The bitumen is introduced at the top of the column, and the column is then eluted successively with solvents of increasing polarity. (A common sequence of solvents is heptane, benzene, and finally methanol.) The most polar organic compounds are adsorbed most strongly to the column packing. A nonpolar solvent like heptane will elute only the nonpolar compounds, which are weakly adsorbed; the heptane fraction will contain the saturated hydrocarbons. Benzene will remove the aromatics and some sulfur compounds. Methanol will elute most of the remainder (resins) except the asphaltenes, which are immobile. It is thus easy to separate four fractions, including the asphaltenes, and to carry out analyses on each of them.

The saturated hydrocarbons are often treated with crystalline urea or synthetic zeolite molecular sieves to separate the *n*-alkanes by urea adduction or molecular sieving. These treatments yield two fractions; one contains the *n*-alkanes, and the other contains the branched and cyclic hydrocarbons.

Most bitumen analyses involve either the whole C_{15^+} bitumen or the saturated hydrocarbons only. Seifert and Moldowan (1978) have recently developed techniques for studying the aromatic fraction.

The analytical method most commonly applied to the saturated hydrocarbon fraction is gas chromatography. This technique is similar to column chromatography, except that the mobile phase is an inert gas rather than a liquid solvent. The column is a very long, thin tube of metal or glass, and it is housed in coiled form in an oven. The rate of movement of the sample through the column depends upon column and sample polarity and column temperature. Often the column temperature is changed during the run according to a fixed program. The most mobile components pass through easily at low temperatures, while the least mobile require higher temperatures in order to move at reasonable rates. Changing column temperature in gas chromatography accomplishes the same thing as changing solvent polarity in column chromatography.

As the individual compounds emerge from the gas chromatograph, they pass through a detector, and their presence is recorded on a chart. In this chromatogram, the area under each peak is proportional to the number of molecules of that particular compound present in the sample. Identification of the individual compounds is accomplished by coinjection of standard compounds.

If urea adduction has been carried out, each of the two saturated hydrocarbon fractions can be analyzed separately. A typical *n*-alkane chromatogram is shown in Figure 6.1. The *n*-alkanes emerge from the chromatograph at more or less regular intervals. A series like this is called a homologous series, and the individual members of the series are called homologs.

Figure 6.2 shows a gas chromatogram for the branched–cyclic fraction of the same oil as that from which the chromatogram of Figure 6.1 was obtained. No homologous series are apparent in Figure 6.2. The urea adduction separation is very useful in producing more easily interpretable chromatograms, because the dominant *n*-alkanes no longer obscure some of the smaller branched–cyclic peaks.

The branched–cyclic fraction is more difficult to analyze, because it contains many kinds of molecules. Combined gas chromatography–mass spectrometry (gc-ms) is very helpful here. The mixture is first run through a gas chromatograph to separate the individual compounds. As they emerge from the gas chromatograph, they automatically enter the mass spectrometer, where they are analyzed individually. Interfacing the mass spectrometer with the gas chromatograph eliminates human handling of the extremely tiny amounts of material emerging from the latter. The mass spectrometer is usually connected to a computer to help handle the huge amounts of data generated.

A mass spectrometer is based on very simple principles. A molecule is ionized by a beam of high-energy electrons. Some of the resultant ions will decompose to give a series of charged fragments of various sizes. These charged fragments are then passed through a magnetic field, where they are separated according to their charge-to-mass ratios. The number of particles of each mass is then recorded, and an output like that shown in Figure 6.3 is obtained. Interpretation of mass spectra is best left to specialists. Much important information on the structure of steranes and triterpanes has emerged from combined gc-ms studies of bitumens and oils.

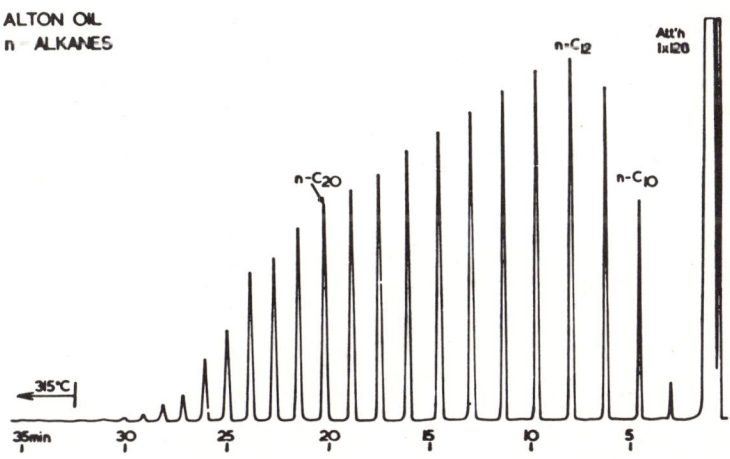

Figure 6.1. Gas chromatogram of *n*-alkanes of an Australian crude oil. (From Mathews et al., 1971; republished with permission of Australian Petroleum Exploration Association, Ltd.)

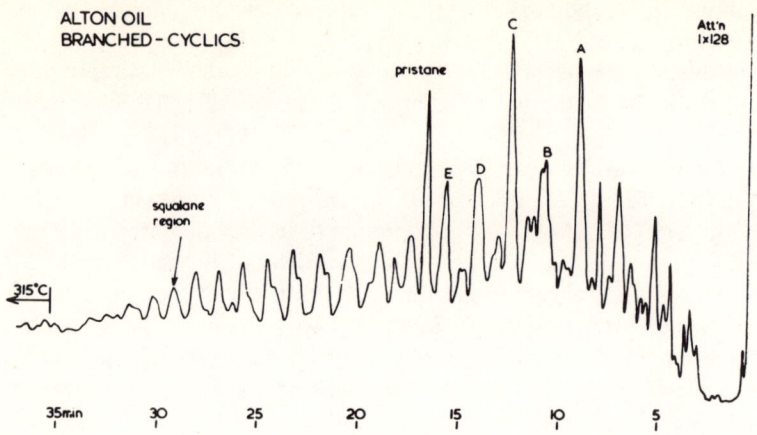

Figure 6.2. Gas chromatogram of the branched and cyclic hydrocarbons of an Australian crude oil. (From Mathews et al., 1971; republished with permission of Australian Petroleum Exploration Association, Ltd.)

Porphyrins have been analyzed in two different ways. Visible and ultraviolet (UV) spectra are able to distinguish between nickel (Ni^{2+}) and vanadyl (VO^{2+}) porphyrins, because UV radiation of different wavelengths is absorbed by the two metal complexes. Mass spectrometry has been capable of much more detailed analyses of porphyrin structures, but because of the complexity of these analyses, the technique has not been applied routinely in exploration.

3.2. Kerogen Analysis

Kerogen analysis can be carried out either on the bitumen-free rock or on unextracted samples, depending upon how a laboratory's analytical flow scheme is set up.

Kerogen is normally isolated from the rock matrix before analysis. The procedures used are derived from standard palynological techniques, and involve removal of carbonate by hydrochloric acid treatment, followed by digestion of silicate minerals by hydrofluoric acid. These procedures produce a kerogen concentrate which also contains other resistant minerals,

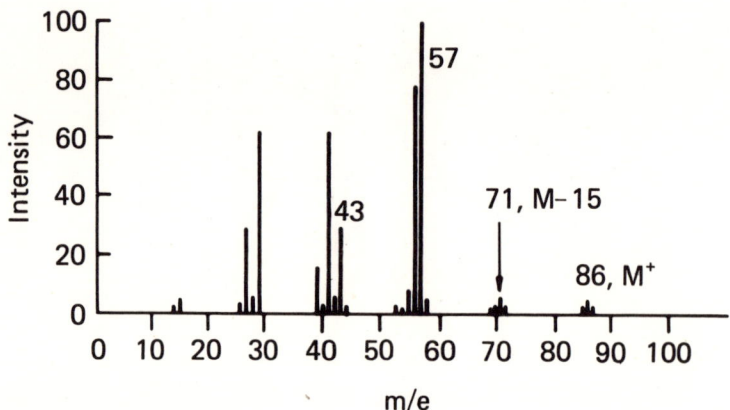

Figure 6.3. Mass spectrum of 3-methylpentane.

especially pyrite. For most purposes, this kerogen concentrate is entirely adequate and is used without further purification. If purification is desired, it can be carried out with at least some success by heavy-liquid separation, centrifugation, or magnetic separation. Kinghorn and Rahman (1980) have also achieved a separation of various kerogen macerals by using density differences.

At this point the kerogen is in a form suitable for microscopic organic analysis (MOA), vitrinite reflectance measurements, or elemental analysis. MOA is carried out on microscope slides prepared as for palynological studies. It is important, however, that no oxidizing or lightening agents be used. The slides can be used to identify the organic macerals present, and to determine the degree of spore darkening or the thermal alteration index (TAI) (Staplin, 1969; Haseldonckx, 1979). Visual kerogen analyses are also important for quality control of data. Anomalous or conflicting data can often be resolved rapidly by an experienced microscopist.

Vitrinite reflectance measurements are carried out with special microscopes. Analysis of fifty to one hundred individual vitrinite particles in each sample is desirable if the results are to have a sound statistical basis. The measured values are plotted in histogram form, and mean values and standard deviations for the data are calculated. Vitrinite is preferred over other macerals because it is present in most samples, regardless of their origin, and because its reflectance characteristics have been well defined by years of work by coal petrologists (Teichmüller, 1971).

Elemental analysis is most conveniently carried out on an automatic carbon–hydrogen–nitrogen analyzer. In these instruments, a small kerogen sample is burned and the amounts of CO_2, H_2O, and N_2 evolved are measured. C, H, and N contents are calculated from the raw data. The main problem with any automated instrument is that it will always give a numerical result, and it sometimes can be difficult to determine whether the instrument is malfunctioning. Close monitoring, careful standardization, and duplicate runs normally eliminate these problems.

If data on the sulfur content of a kerogen sample are required, either the pyrite (FeS_2) must first be removed, or else a compensating calculation must be made. Because of these experimental difficulties, sulfur contents of kerogens are not normally measured. Oxygen is best estimated by CO_2 yield during pyrolysis in the Rock-Eval, which is discussed next. next.

The Rock-Eval, an ingenious instrument designed specifically for source-rock evaluation, has recently come on the market. It analyzes both bitumen and kerogen automatically in about twenty minutes, and is intended to give a complete analysis of oil-source capability and thermal maturity of the kerogen. A complete description of this method is given by Espitalié et al. (1977) and by Clementz et al. (1979).

In a Rock-Eval analysis, a small (100 mg) sample of powdered rock is heated in the absence of oxygen. In the early, low-temperature heating phase the bitumen molecules are mobilized, resulting in a thermal extraction of the bitumen already present in the rock. The bitumen obtained by thermal extraction is approximately equivalent to that obtained by solvent extraction. The thermally extracted bitumen (S_1) is then passed through a gas chromatograph detector and measured quantitatively. No kerogen decomposition occurs at this stage.

The temperature in the oven then is increased to bring about breakdown (pyrolysis) of the kerogen. This process, analogous to natural kerogen catagenesis, results in the formation of new bitumen molecules. These mobile products (S_2) are also passed through the gc detector and analyzed.

Inorganic gases are also produced during kerogen pyrolysis; one of these gases, CO_2, is also collected and measured. Care is taken to insure that no decomposition of inorganic carbonates occurs during the CO_2 collection phase.

The S_1 fraction represents all of the diagenetic and catagenetic bitumen present in the rock before it was brought to the lab. As such it is an indicator of the amount of bitumen generation that has already occurred. The S_2 fraction represents the remaining oil-generative capacity in the rock at the time of sampling. The ratio $S_1/(S_1+S_2)$, called the transformation ratio, is a measure of the degree of catagenesis which has occurred.

Carbon dioxide is used as an indirect measure of the oxygen content of the kerogen, while the S_2 fraction is related to the hydrogen content. If the organic carbon content of the rock is known from independent measurements, and if standards have been run to calibrate the detector response of the gas chromatograph, an estimate of both the hydrogen and oxygen contents of the kerogen can be obtained. These values are termed the hydrogen index and the oxygen index, respectively (Espitalié et al., 1977).

Plotting the hydrogen index against the oxygen index results in a graph of the form shown in Figure 6.4. This figure is analogous to the earlier Tissot and van Krevelen diagrams (Figure 3.1). Rock-Eval pyrolysis can therefore be employed to determine kerogen type.

Finally, the temperature at which pyrolysis yields the maximum amount of bitumen is also related to the thermal maturity of the kerogen. More mature kerogens require higher temperatures for their pyrolysis because the weaker chemical bonds have already been broken.

The Rock-Eval is capable of providing a large amount of data about the oil-source history and future of a rock. It is rapid and requires little sample. Application of the Rock-Eval to source-rock studies is rather new, however, and there is as yet no single standard approach to using the instrument and the data it generates. Some possible applications of the method to the solution of source-rock problems are discussed in Chapter 7.

Carbon isotope ratios have been used for many years in organic geochemistry. There are two stable isotopes of carbon: carbon-12 (^{12}C) and carbon-13 (^{13}C). The ^{13}C content of carbon

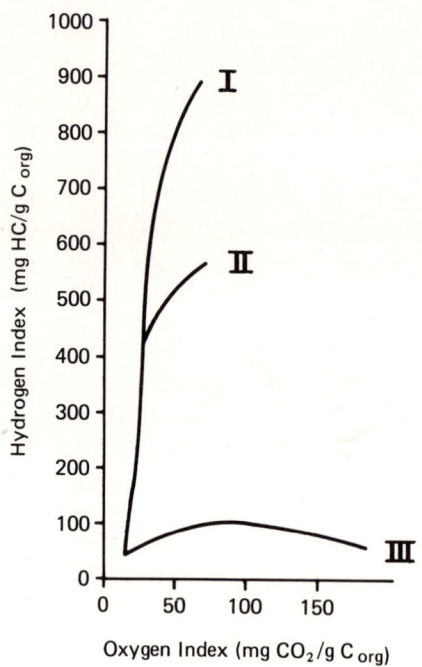

Figure 6.4. Plot of hydrogen index versus oxygen index for different kerogen types run on the Rock-Eval. (From Espitalié et al., 1977; republished with permission of Institut Francais du Pétrole)

in nature is about 1.1%, but significant deviations occur, depending upon the composition and history of a particular carbonaceous material. The range of actual ^{13}C contents in nature is from 1.02% to 1.13% (Fuex, 1977). Because variations in $^{13}C/^{12}C$ ratios are quite small, the ratios must be measured with great accuracy if one is to derive any meaningful data from them.

Carbon isotope ratios are determined by completely oxidizing (burning) the substance to be analyzed to CO_2, and then measuring the relative amounts of $^{13}CO_2$ (mass = 45 amu) and $^{12}CO_2$ (mass = 44 amu) with a special mass spectrometer. The measured values are always compared to the ratio for a standard substance, and are reported as $\delta^{13}C$ values relative to that particular standard. $\delta^{13}C$ is defined by the Equation 6.2; its values are reported in "permil," abbreviated "°/oo."

$$\delta^{13}C = \left[\frac{(^{13}C/^{12}C) \text{ sample}}{(^{13}C/^{12}C) \text{ standard}} - 1 \right] \times 1000 \quad (6.2)$$

The most common standard is a carbonate belemnites fossil from the PeeDee Formation of South Carolina, abbreviated PDB. A few groups report values relative to a National Bureau of Standards oil standard, NBS-22.

The usual precision attainable in $\delta^{13}C$ measurements is ±0.1 or 0.2 °/oo, but if higher precision is required, it can be obtained with more effort.

Carbon isotope measurements can be applied to any material that contains carbon. Natural gas, petroleum, bitumen, kerogen, and coal have all been analyzed many times, and a large library of reference data exists.

7 DATA INTERPRETATION

1. Introduction

Before the techniques mentioned in Chapter 6 are applied for interpretative purposes, the investigator must decide which of the four interpretation questions listed below is most appropriate.

1. What is the total oil-source potential of a rock unit, and how much of that oil has already been generated?
2. What is the total gas potential of a rock unit, and how much of that gas has already been generated?
3. What type of growth environment produced the sedimented organic material, and under what conditions was the material transported and deposited?
4. Is there a correlation between two different samples of organic material? For example, did a particular oil come from a certain source rock?

Any particular interpretation problem may require an answer to one or more of these questions. Some analytical data may be helpful in answering one of these questions, but not at all useful in dealing with the others. Table 7.1 summarizes the uses of each type of data relative to the interpretative questions.

2. Aspects of the Oil-Source Potential Problem

The first question, which deals with source potential, is generally of greatest interest. A detailed discussion of the philosophy and methodology of source rock analysis is useful at this point; the ideas in a recent paper (Waples, 1979) will form the basis of the discussion.

There are two distinct aspects to the problem of source capacity. The first is, "How much total oil could be generated from a given volume of source bed if generation went to completion?," and the second is, "How much oil has already been generated from a given volume of the source bed?" The answer to the former corresponds to what may be called "Total Oil," while the answer to the latter can be called "Oil Already Generated." Both aspects are important, but in a given geologic setting, one may be much more relevant than the other. It is essential, however, that "Total Oil" and "Oil Already Generated" not be confused.

The factors which govern oil-source capacity of a sedimentary rock are the quantity of organic material in the rock, the oil-generative quality of that organic matter, the thermal maturity of the kerogen, and the expulsion efficiency of the bitumen from the rock.

Table 7.1. Analytical Techniques and their Proper Application for Interpretation

Type of Analysis	Questions Answered[a]
Total organic carbon	1,2
Bitumen-free organic carbon	1,2
Total bitumen quantity	1
n-paraffins	1, 2, 3, 4
isoprenoids	3, 4
steranes and triterpanes	3, 4
carbon isotopes	3, 4
porphyrins	4
elemental kerogen analysis	1, 2, 3
TAI, vitrinite reflectance	1, 2
microscopic organic analysis	1, 2, 3
Rock-Eval	1, 2, 3

[a] 1. Total oil = source potential
2. Total gas = source potential
3. Growth environment
4. Correlations

There are two distinct approaches to the problem of calculating "Total Oil" and "Oil Already Generated." One of these we might call the direct approach: the rock is actually subjected to catagenetic conditions in order to measure directly how much bitumen is produced. The Rock-Eval and other pyrolysis techniques are the methods of choice for the direct approach to source-rock evaluation.

Indirect approaches involve measuring quantity, quality, and maturity of the organic matter independently, and then combining these data to predict what the oil generative

Table 7.2. Kerogen and Bitumen Parameters as Applied to Source-Potential Evaluation

Material analyzed	Analysis	Quantity	Quality	Maturity
Kerogen & bitumen together	Total organic carbon (TOC)	•		
Kerogen & bitumen together	Rock-Eval (pyrolysis)	——— Direct Method ———		
Kerogen	Bitumen-free organic carbon (BFOC)	•		
Kerogen	H/C ratio		•	•
Kerogen	Thermal Alteration Index (TAI)			•
Kerogen	Vitrinite reflectance (R_O)			•
Kerogen	Microscopic Organic Analysis (MOA)		•	
Kerogen	Electron Spin Resonance (esr)			•
Kerogen	Fluorescence		•	•
Bitumen	Bit./BFOC			•
Bitumen	HC/Total bit.			•
Bitumen	n-paraffins			•

capacity should be. The indirect methods all rely on the assumption that it is possible to predict bitumen generative capacity from the chemical and physical properties of a kerogen.

Many different parameters have been used by geochemists as measures of quantity, quality, and maturity. Table 7.2 lists some of these and indicates which aspects of the source-potential question are dealt with by each.

3. Direct Method for Measuring Oil-Generative Capacity

As is shown in Figure 6.4, the hydrogen index obtained with the Rock-Eval is expressed as milligrams of hydrocarbon per gram of organic carbon present in the rock. It is therefore easy to convert the hydrogen index to other more familiar units, such as barrels per cubic mile or gallons per ton of rock.

We should note parenthetically that it is not actually necessary to measure the organic carbon content of the rock in order to use the Rock-Eval. Because it is the whole powdered rock which is pyrolyzed, the simplest approach is to relate the weight of hydrocarbons generated directly to the weight of rock. Further conversion to barrels per cubic mile is then very easy.

The hydrogen index is a direct measure of the hydrocarbon-generative capacity of the kerogen in the rock. If we then can make estimates of total rock volume and its thermal maturity, and of expulsion, migration, and accumulation efficiencies, we will have a means of estimating oil in place.

It is important to realize that the hydrocarbons generated by pyrolysis contain both gaseous and liquid species. The relative proportions of gas and oil will depend upon the kerogen type. Type I kerogens may give as much as 80% oil, while Type III kerogens may only yield 10% oil and 90% gas. Any calculation of oil-generative capacity must take into consideration the fact that the Rock-Eval detector lumps all hydrocarbons together, regardless of molecular size.

To show how the "Total Oil" can be calculated by using the Rock-Eval, consider as an example an "average" shale. This rock contains about 1% organic carbon and is capable of generating about 300 mg of hydrocarbons per gram of organic carbon. Let us assume that 40% of these hydrocarbons are liquid and the rest gas. If we further assume plausible values of 2.3 g/cm^3 for shale density and 0.9 g/cm^3 for oil density, it is easy to show that this shale will generate about 80 million barrels of liquid hydrocarbons per cubic mile when full maturity is attained. A richer shale, say one containing 5% organic carbon, that consisted of the same kind of kerogen would generate 400 million barrels per cubic mile. These are very large numbers, but they pale by comparison with the calculated generation in a cubic mile of Green River Shale (assumed to be fully mature, yielding 600 mg of liquid hydrocarbons per gram of organic carbon, and having a TOC value of 20%): 8 billion barrels per cubic mile.

The magnitudes of these numbers, even those for an average shale, suggest that oil generation in general may not be a problem. In many regions the important factors which determine whether the region will host large reservoirs of petroleum may be expulsion, migration, and accumulation efficiency, rather than generative capacity.

There are some limitations to the direct (pyrolysis) methods of determining oil-source capacity. As has already been mentioned, no distinction is made between gaseous and liquid products. It is therefore useful to obtain some appraisal of the kerogen type so that a reasonable estimate of the amount of each type of product can be made. If the oxygen index is obtained, the Rock-Eval pyrolysis and a TOC measurement are sufficient to identify kerogen type. Other analyses, such as H/C ratios or maceral analyses, can also provide much information about kerogen type.

Another limitation of Rock-Eval analyses is that they can be applied directly only when the kerogen being pyrolyzed is immature. The hydrogen index is lessened by an increase in maturity, so unduly low estimates of "Total.Oil" would be obtained from pyrolysis of mature samples unless a correction for maturity were made. There are two ways of avoiding this problem: work only with immature samples, or make a correction for maturity in a manner similar to that suggested for H/C ratios later in this chapter.

4. Indirect Methods for Estimating Oil-Generative Capacity

Methods in which kerogen analyses are employed are necessarily indirect methods because the hydrocarbon products are inferred rather than observed. In any indirect approach, it is necessary to evaluate quantity, quality, and maturity, and then to combine them to give an overall evaluation of oil-generative capacity. Each of these factors is considered individually below.

4.1 QUANTITY

Many workers (e.g., Hunt, 1972; Welte, 1965) have been concerned with the concentration of organic carbon in source beds and with the implications of these concentrations for oil-source capacity. Other things being equal, the quantity of oil which is generated in a given volume of source bed is linearly related to the source bed's organic carbon content (Cook, 1974). We may therefore define a *quantity factor* as the organic carbon content of the source bed. Organic carbon content can easily be measured as total organic carbon or bitumen-free organic carbon (see Table 7.2 and Chapter 6). Because the organic carbon content of an average shale is about 1% (Hunt, 1972), the quantity factor for an average shale is 1.0.

4.2 QUALITY

A number of parameters commonly are used to characterize the oil-source quality of kerogens (Table 7.2). One is based exclusively on visual kerogen analysis in transmitted light. Results were originally reported as "% Amorphous" material, but more recently coal maceral terminology (Figure 3.1) has been applied to kerogen studies (Dow, 1977). The maceral groups alginite and exinite are often reported as "%Alginite + Liptinite." Such evaluations are subjective, but nevertheless useful, in characterizing kerogens.

A plausible simplifying assumption is that alginite and exinite generate oil, and that vitrinite and inertinite do not. The relationship between kerogen type and the quality factor is therefore simple and linear. The *quality factor* is defined to be 1.0 for a kerogen of approximately average quality (50% alginite + liptinite, determined visually). It is not necessary, of course, to assume that only alginite and exinite generate oil, or that they are equally productive. Several other plausible but arbitrary assumptions about relative oil-generative efficiencies, such as those given in Table 7.3, could be justified. The various models presented are all rather similar, and the data presently available do not favor one model over the others.

Another method of defining kerogen quality involves measuring its atomic H/C ratio. Because kerogen H/C ratios are functions of both kerogen quality and thermal maturity, the maturity component must be removed before H/C ratios can be used as a quality parameter (cf. Dow, 1977; Tissot et al., 1974). It is therefore necessary to retrace the kerogen's maturation pathway in order to determine its H/C ratio in the original immature state. It is this immature H/C ratio which defines kerogen quality.

The H/C ratio that a kerogen had when it was thermally immature can be estimated if both its present H/C ratio and thermal maturity are known. The kerogen is located on a graph of H/C ratio versus vitrinite reflectance. Such a plot is philosophically very similar to a Tissot or van Krevelen diagram (Figure 3.1), except that vitrinite reflectance has replaced the O/C ratio.

Table 7.3. Several Possible Assumptions about the Relative Oil-Generative Capacity of Different Macerals

| | Relative oil generative capacity by maceral type | | | |
Model	I Alginite	II Exinite	III Vitrinite	IV Inertinite
A	1	1	0	0
B	1	0.9	0.1	0
C	1	0.8	0.1	0
D	1	0.7	0.1	0

Point P in Figure 7.1 represents such a kerogen. When it was immature, this kerogen had the H/C ratio represented by point A. It has already progressed down the maturation pathway to P, and with additional maturation will continue toward B. Once a kerogen has been located on the H/C diagram its maturation pathway can be interpolated and its immature H/C ratio determined. The quality factor is related to the immature H/C ratio in the manner shown in Figure 7.2. H/C ratios of immature kerogens vary from about 0.4 (characteristic of some inertinites) to 1.6 (about maximum for pure alginite). The H/C ratio for an average immature kerogen is about 1.0 (Waples, 1977); this H/C ratio is therefore assigned a quality factor of 1.0. Atomic H/C ratios can be directly compared with maceral type data by correlating quality factors obtained by the different methods.

Pyrolysis can also be used as a measure of kerogen quality. It is, of course, necessary either that all samples be thermally immature or that the pyrolysis yield for a mature sample be

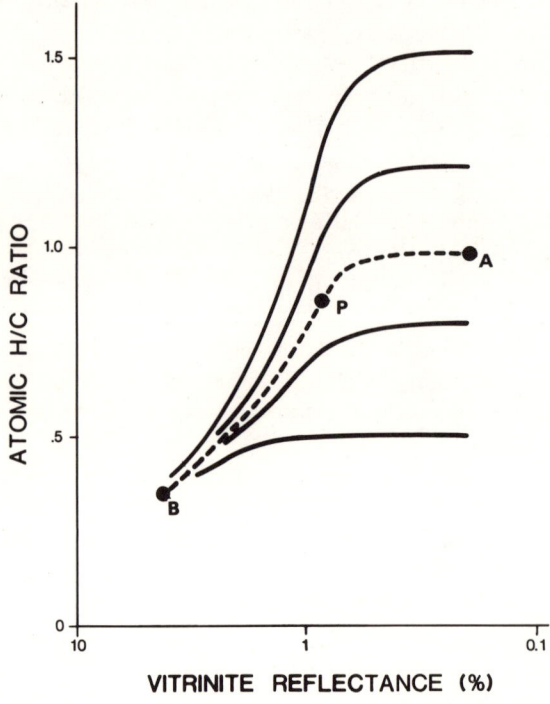

Figure 7.1. Calculations of the immature kerogen H/C ratio from present-day H/C ratio and vitrinite reflectance data. (From Waples, 1979, republished with permission of The American Association of Petroleum Geologists)

corrected for the effects of maturity. Figure 7.3 shows one possible way of scaling the measured pyrolysis values. The nonlinear shape of the curve reflects the varying proportions of gaseous and liquid hydrocarbons in the pyrolysis products of different kerogens.

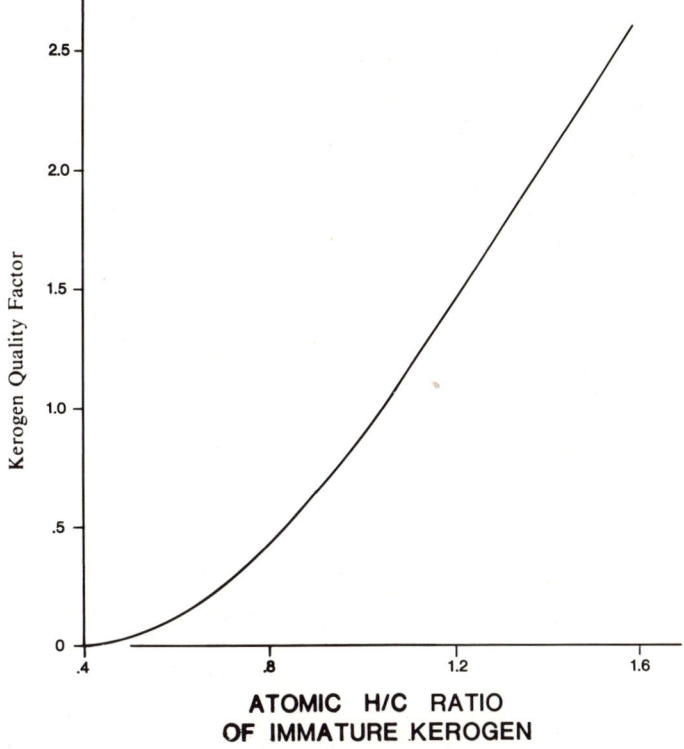

Figure 7.2. Kerogen quality factor as a function of H/C ratio of the immature kerogen. (From Waples, 1979; republished with permission of The American Association of Petroleum Geologists)

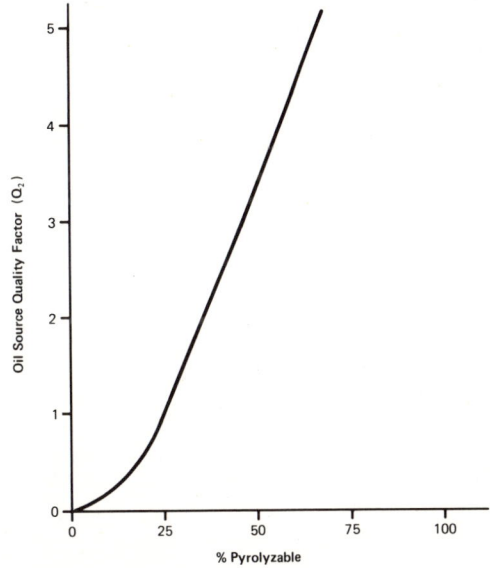

Figure 7.3. Kerogen oil-source quality factor as a function of pyrolysis yield. "% Pyrolyzable" refers to the percent of carbon in kerogen that is lost during pyrolysis.

4.3. MATURITY

The two most commonly used parameters for measuring the thermal maturity of kerogens are thermal alteration index (TAI) and vitrinite reflectance (R_o). Both techniques are usually valid, but reflectance work is generally considered to be less subjective and more precise.

TAI is measured on a color scale defined by Staplin (1969); the scale ranges from 1 to 5, with higher values indicating greater maturity. The oil-generative zone is thought to correspond approximately to the golden-brown through brown stages. The exact numerical values attached to each color vary from laboratory to laboratory.

A detailed but arbitrary (in the sense that this is one laboratory's scheme) conversion table is given in Table 7.4; it will be used extensively in the practice problems (Chapter 9). On this scale, the TAI values of oil-generative zones are approximately in the range of 2.6 to 3.2. TAI values are sometimes reported as a range of values (e.g., 2–2.5), indicating an uncertainty on the part of the microscopist. When an entire profile is studied, however, the presence of a few uncertain samples usually is not too troublesome.

Vitrinite reflectance measurements record the percentage of incident light which is reflected by individual vitrinite particles. A large number of particles is analyzed in each sample, and the results are reported in histogram form, with an accompanying statistical analysis of the histogram's shape. Reflectance measurements ideally would produce histograms approximating the normal curve. Good histograms, such as the one shown in Figure 7.4A, have a single mode and a small standard deviation. Poor histograms, such as

Table 7.4. Conversion between Thermal Alteration Index (TAI) and Vitrinite Reflectance (Ro)[a,b]

R_O	TAI	R_O	TAI
0.30	2.0	1.26	3.15
0.34	2.1	1.30	3.2
0.38	2.2	1.33	3.25
0.40	2.25	1.36	3.3
0.42	2.3	1.39	3.35
0.44	2.35	1.42	3.4
0.46	2.4	1.46	3.45
0.48	2.45	1.50	2.5
0.50	2.5	1.62	3.55
0.55	2.55	1.75	3.6
0.60	2.6	1.87	3.65
0.65	2.65	2.0	3.7
0.70	2.7	2.25	3.75
0.77	2.75	2.5	3.8
0.85	2.8	2.75	3.85
0.93	2.85	3.0	3.9
1.00	2.9	3.25	3.95
1.07	2.95	3.5	4.0
1.15	3.0	4.0	4.0
1.19	3.05	4.5	4.0
1.22	3.1	5.0	4.0

[a] After Waples, 1980

[b] The exact TAI value that corresponds to a given R_O value depends upon the particular laboratory's system for assigning TAI values to spore colors.

those shown in Figures 7.4B and C, are broad, bimodal, or diffuse. Much of the scatter in such histograms may be due to the presence of reworked material, misidentified nonvitrinite material, or contamination. R_O values less than approximately 0.5% are subject to uncertainty because the organic material being analyzed has not yet become true vitrinite.

Reflectance values should always be reported along with the accompanying histograms (or at the very least a statistical summary of the distribution). If such information is not included, the interpreter has no idea of the confidence level of the data. Unfortunately, such omissions have been common in the past.

If it is assumed that oil generation follows first-order chemical kinetics (see Chapter 8), then a plot of generation efficiency versus R_O can be constructed (Figure 7.5). Many workers (e.g., Dow, 1977) define the principal zone of oil formation to lie between vitrinite reflectance values of 0.6% and 1.2%. These limits are not the absolute beginning and end of all oil generation, however; chemical reaction rate theory requires that some oil formation occur earlier and some later. It seems reasonable to assume that the principal zone of oil formation represents 90% of all oil generation. (This assumption appears reasonable in light of present knowledge, but it should not be regarded as absolutely correct.) Figure 7.5 indicates that about

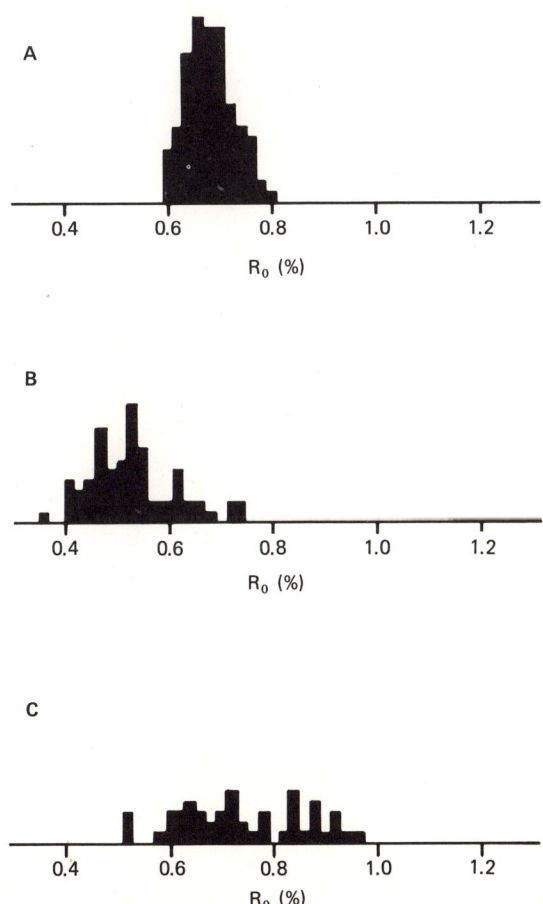

Figure 7.4. Vitrinite reflectance histograms for three samples. A, Good histogram; B and C, Poor histograms of little value to the geochemist.

half of the total possible oil generation has occurred when $R_o = 0.9$. Generation efficiency does not refer to the absolute quantity of oil generated, however, but rather to the degree to which the possible oil generation has actually occurred.

A similar relationship between TAI and the maturity factor can be constructed easily by using Figure 7.5 and Table 7.4. This exercise is left to the reader and appears in some of the problems in Chapter 9.

Various bitumen parameters are also related to thermal maturity. In theory, these could also be used to estimate the degree of catagenesis in a source bed, but in practice it is very difficult to obtain the same precision and accuracy from bitumen data that can be obtained from TAI or reflectance data. Difficulties with bitumen analyses often stem from contamination (Chapter 6) or migration (Chapter 5).

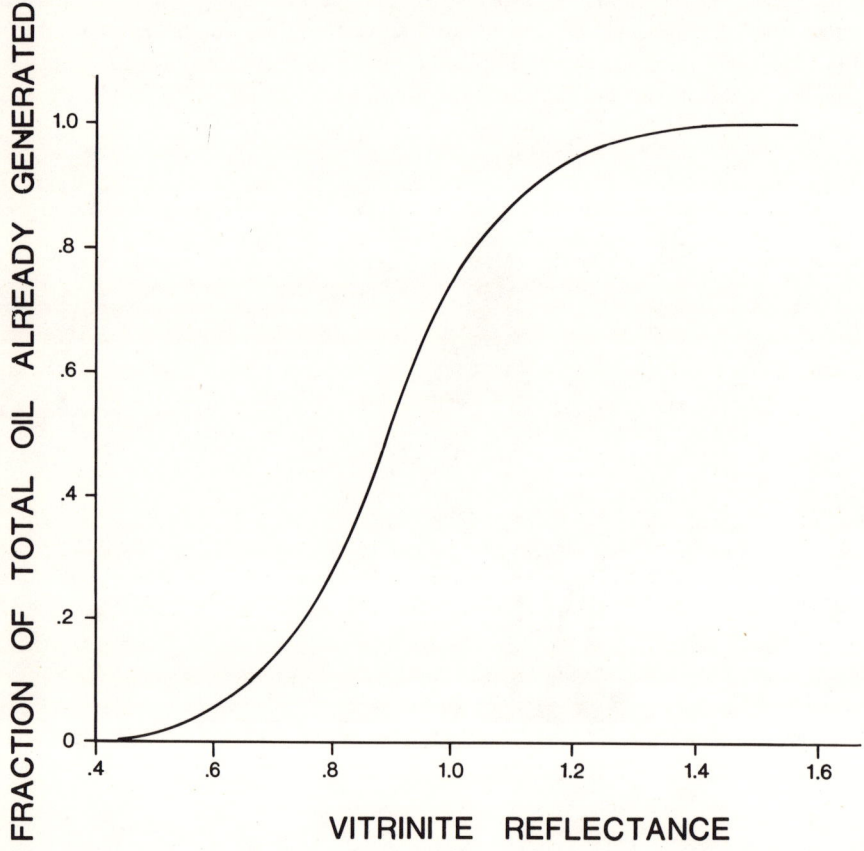

Figure 7.5. Oil generation efficiency as a function of kerogen maturity as measured by vitrinite reflectance, based on assumptions outlined in the text. (From Waples, 1979; republished with permission of The American Association of Petroleum Geologists)

4.4 EXPULSION EFFICIENCY

Expulsion efficiency is poorly understood, but must be of significance in source-bed evaluation. Welte (1965) and others have noted that a minimum organic carbon content (0.5%) in a source bed is necessary before bitumen expulsion can occur, even though small amounts of bitumen may have been generated by thermal cracking. Welte's reasoning is that bitumen is adsorbed on kerogen and clay surfaces, and that until these adsorption sites are filled, no bitumen can be expelled from the source bed.

An expulsion efficiency factor should properly be included, therefore, in "Total Oil" calculations. To derive satisfactory expulsion efficiency factors, several questions must be answered. These are:

- How efficient is bitumen expulsion from source beds of very low organic richness?
- How does expulsion efficiency depend on organic carbon content for typical source beds?
- Is expulsion efficiency lowered when the source bed matrix is essentially organic, as in the case of coals or oil shales?
- How does source bed lithology affect expulsion efficiency? Are carbonates significantly different from shales?
- Should porosity and permeability be treated as independent factors affecting expulsion efficiency?
- How important is source bed thickness to expulsion efficiency?

Many of these questions have already been discussed in Chapter 5, but a much better understanding of migration is required if they are to be answered intelligently. It is preferable at the present time, therefore, to leave expulsion efficiency entirely out of source-rock calculations and to bring it in later on an *ad hoc* case-by-case basis.

5. Evaluation of Oil-Source Capacity

Four independent conditions must be satisfied by sedimentary organic material before petroleum can be generated and expelled. There must be a sufficient quantity of organic material, it must be of oil-generating quality, it must have experienced a sufficient amount of thermal maturation, and the bitumen formed must have been expelled. These conditions impose simultaneous requirements on the system; if any one condition is not met, no migratable oil can be formed. Furthermore, if one condition is met only partially, oil generation will be severely reduced.

Because the quantity of petroleum actually generated in and expelled from a source bed depends on four simultaneous conditions, the system can be described by the equation

$$\text{Oil expelled} = (Q_1)(Q_2)(M)(E) \tag{7.1}$$

where Q_1 and Q_2 are the scaled factors for quantity and quality, respectively, M is the percentage generation factor corresponding to the thermal maturity of the kerogen, and E is the expulsion efficiency of the source sequence. As mentioned previously, however, expulsion efficiency is not quantifiable at the present time, so it is preferable to omit the E factor completely. Equation (7.1) thus reduces to

$$\text{Oil Already Generated} = (Q_1)(Q_2)(M) \tag{7.2}$$

A calculation of "Total Oil" assumes that thermal maturation is complete, and that M therefore always equals 1.0 (Figure 7.5). Equation (7.2) thus becomes

$$\text{Total Oil} = (Q_1)(Q_2) \tag{7.3}$$

Illustrative data for hypothetical kerogens of varying quality and quantity, and the calculated "Total Oil" values of the kerogens, are shown in Table 7.5. All examples have "Total Oil" values between 0.05 and 36, a range of nearly three orders of magnitude. Thus, the Green River Shale ("Total Oil" = 36) should have nearly 1000 times the total oil-source capacity of an organic-lean, poor quality kerogen ("Total Oil" = 0.05). Because the quantity and quality factors of an average rock were normalized to 1.0, a "Total Oil" value of 1.0 represents an "average" shale. As was pointed out earlier in this chapter, an "average" shale generates 80 million barrels of oil per cubic mile of rock. We can therefore say that a "Total Oil" value of 1.0 corresponds to about 80 million barrels of bitumen per cubic mile of source rock.

Some uncertainties emerge when we try to correlate the direct method with indirect ones. For example, the Green River Shale evaluated by pyrolysis was calculated to have a "Total Oil" value of 8 billion barrels per cubic mile, but the prediction of the indirect approach is less than 3 billion (Table 7.5). This discrepancy appears to be serious, but in actuality involves only an uncertainty of a factor of 3 in the estimate. Most of this uncertainty can probably be ascribed to the assumption about the uniform oil-generative capacity of all alginites and liptinites, and suggests that further refinement of the Q_2 scale by correlation with pyrolysis would be wise.

Table 7.5. "Total Oil" Values for Selected Kerogens [a,b]

	Raw Data		Scaled Values		
% Corg	% Alg. + Lipt.	Q_1	Q_2		"Total Oil" [c]
.1	50	0.1	1.0	0.1	(8)
.1	100	0.1	2.0	0.2	(16)
.5	5	0.5	0.1	0.05	(4)
.5	100	0.5	2.0	1.0	(80)
1.0	25	1.0	0.5	0.5	(40)
1.0 [d]	50	1.0	1.0	1	(80)
1.0	100	1.0	2.0	2	(160)
1.5	25	1.5	0.5	0.75	(60)
1.5	50	1.5	1.0	1.5	(120)
1.5	100	1.5	2.0	3	(240)
2.0	25	2.0	0.5	1	(80)
2.0	100	2.0	2.0	4	(320)
5.0	25	5.0	0.5	2.5	(200)
5.0	100	5.0	2.0	10	(800)
10.0	10	10.0	0.2	2	(160)
10.0	100	10.0	2.0	20	(1600)
20.0	90	20.0	1.8	36	(2880)
50.0	5	50.0	0.1	5	(400)
50.0	20	50.0	0.1	20	(1600)

[a] $M = 1.0$
[b] From Waples (1979)
[c] Values in parentheses are millions of bbl/mi³.
[d] "Average" source bed.

Table 7.6. "Oil Already Generated" Values for Selected Kerogens [a]

% Corg	% Alg. + Lipt.	R_0	Q_1	Q_2	M	"Total Oil"	"Oil Already Generated" [b]	
1.0	50	0.50	1.0	1.0	2.0	1	0.02	(1.6)
1.0	50	0.70	1.0	1.0	0.11	1	0.11	(8.8)
1.0	50	0.90	1.0	1.0	0.30	1	0.30	(24)
1.0	50	1.10	1.0	1.0	0.72	1	0.72	(58.0)
1.0	50	1.30	1.0	1.0	0.95	1	0.95	(76)
5.0	25	0.50	5.0	0.5	0.02	2.5	0.05	(4)
5.0	25	0.70	5.0	0.5	0.11	2.5	0.28	(22.0)
5.0	25	0.90	5.0	0.5	0.30	2.5	0.75	(60)
5.0	25	1.10	5.0	0.5	0.72	2.5	1.80	(144)
5.0	25	1.30	5.0	0.5	0.95	2.5	2.38	(190)
20.0	90	0.40	20.0	1.8	0.00	36	0	(0)
20.0	90	0.50	20.0	1.8	0.00	36	0.72	(58.0)
20.0	90	0.60	20.0	1.8	0.05	36	1.80	(144)
20.0	90	0.70	20.0	1.8	0.11	36	3.96	(317)
20.0	90	0.80	20.0	1.8	0.18	36	6.48	(518)

[a] After Waples (1979)
[b] Values in parentheses are millions of bbl/mi³.

"Oil Already Generated" is calculated by substituting the actual M value into equation (7.2). "Oil Already Generated" values from a few hypothetical kerogens at various stages of thermal maturation, and their corresponding "Total Oil" values, are shown in Table 7.6.

Because "Total Oil" and "Oil Already Generated" values are functions only of oil generation, and not of expulsion, they are of limited value in exploration applications. As discussed in Chapter 5, expulsion and migration from a source sequence are generally considered to be very inefficient and highly variable processes which could alter greatly any specific volumetric predictions based on "Oil Already Generated" values. Nevertheless, "Oil Already Generated" and "Total Oil" values are very useful in a relative sense for comparing source beds with each other in a semiquantitative way. If in the future it becomes possible to make reasonable estimates of migrational efficiency, source-rock analysis will become extremely powerful.

6. Application to Source Rock Analysis

6.1 C.O.S.T. B-2 Well

Figure 7.6 shows the quantity, quality, and maturity profiles in the Continental Offshore Stratigraphic Test (C.O.S.T.) B-2 Well in the Baltimore Canyon area. Quantity was measured as TOC, quality was determined by measuring the present-day kerogen H/C ratios and extrapolating back to the corresponding immature H/C ratios, and maturity was based on the best straight-line fit to the vitrinite reflectance data (individual points not shown). Each sample was then numerically evaluated according to equations (7.2) and (7.3), and the resultant "Total Oil" and "Oil Already Generated" values were plotted (Figure 7.7). The numerical approach and the visual presentation simplify interpretation of the raw data.

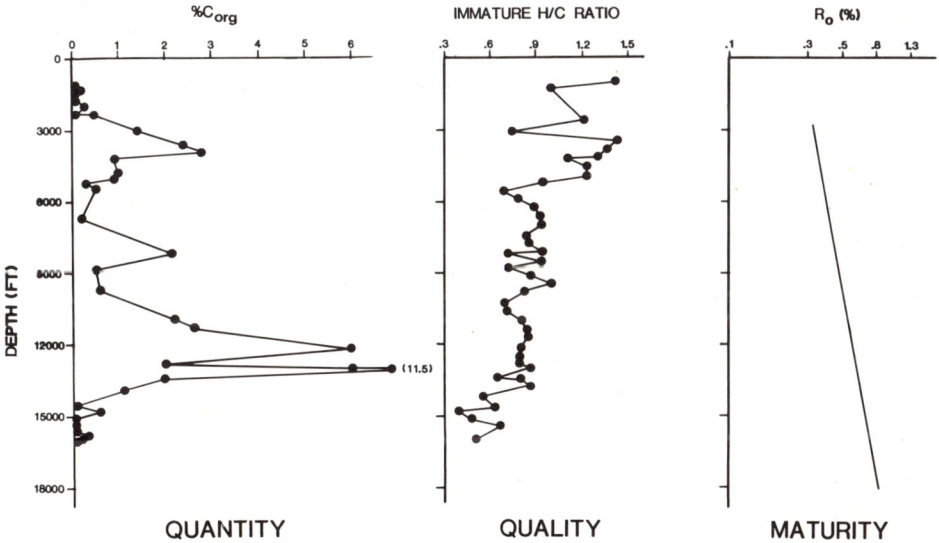

Figure 7.6. Quantity, quality, and maturity profiles for the C.O.S.T. B-2 well. (From Waples, 1979; republished with permission of The American Association of Petroleum Geologists)

The "Total Oil" plot in Figure 7.7 shows two regions of relatively high oil-source capacity: 2500–5000 feet and 11,000–13,500 feet. Most of the samples in these portions of the section have "Total Oil" values greater than 1.0, and some are as high as 6.

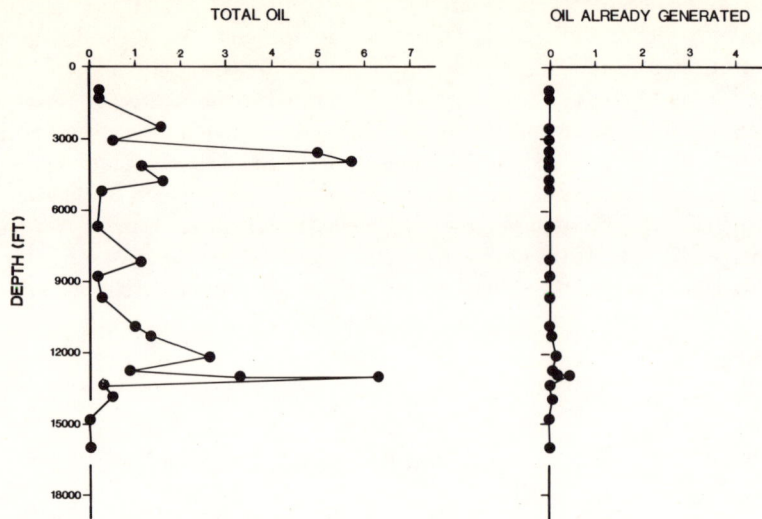

Figure 7.7. "Total Oil" and "Oil Already Generated" profiles for the C.O.S.T. B-2 well. (From Waples, 1979; republished with permission of The American Association of Petroleum Geologists)

It is equally obvious from the "Oil Already Generated" curve in Figure 7.7, however, that because thermal maturity in the section is low, very little oil has actually been generated. These conclusions are in excellent agreement with those of Claypool et al. (1977).

On the basis of a quick analysis of Figure 7.7, one could conclude that although the section studied has itself not generated any significant quantities of petroleum, the two above-mentioned sequences could be excellent source rocks under the proper geologic conditions. Attention should therefore be shifted to locating other geologic environments where these potential oil sources have been exposed to the proper maturation conditions for oil generation. Application of the time–temperature criteria of Lopatin (1971) (Chapter 8) would be helpful in predicting where the proper maturation conditions might exist.

6.2 C.O.S.T. #1 Well

The C.O.S.T. #1 Well, drilled off the Texas Gulf Coast, represents a typical example from that province. Figure 7.8 shows the quantity, quality, and maturity profiles for the section penetrated. The data were taken from the source rock study conducted by Geochem Laboratories, Inc. Organic carbon contents hover between 0.5% and 1.0%. The section is therefore organically rather lean. Kerogen quality, as measured by microscopic organic analysis, is somewhat variable, but it generally emerges as about average or slightly above. The section is thermally immature down to about 12,000 feet, below which the oil-generative zone is reached. Vitrinite reflectance values at total depth (about 16,000 feet) are approximately 1.1%, indicating that peak generation has been attained.

Figure 7.9 presents the oil-generative history and future of the section. "Oil Already Generated" is quite low throughout the section. Maximum values of about 0.4 (compared to 1.0 for a fully mature, average shale, or 0.5 for an average shale at peak generation) are encountered around 14,500–15,000 feet. These results suggest that the C.O.S.T. #1 Well represents a significantly below-average prospect as a present-day oil-source rock.

"Total Oil" is also meagre. There is little hope that this section, even if exposed to optimal thermal maturation, would be a prolific source. The maximum "Total Oil" values are about 1, and are thus comparable to those for an average shale. Much of the section has a "Total Oil"

Data Interpretation   79

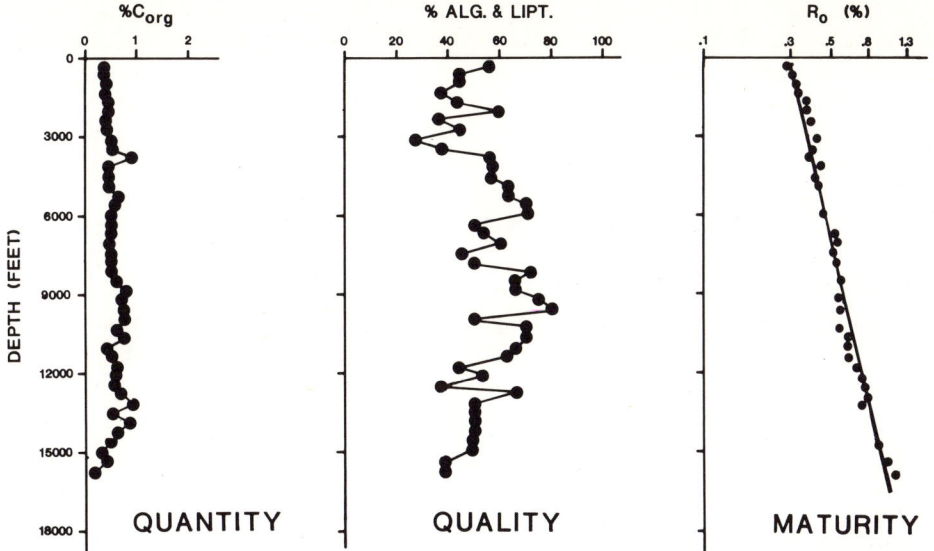

Figure 7.8. Kerogen quantity, quality, and maturity data for the C.O.S.T. #1 Well, offshore Texas Gulf Coast.

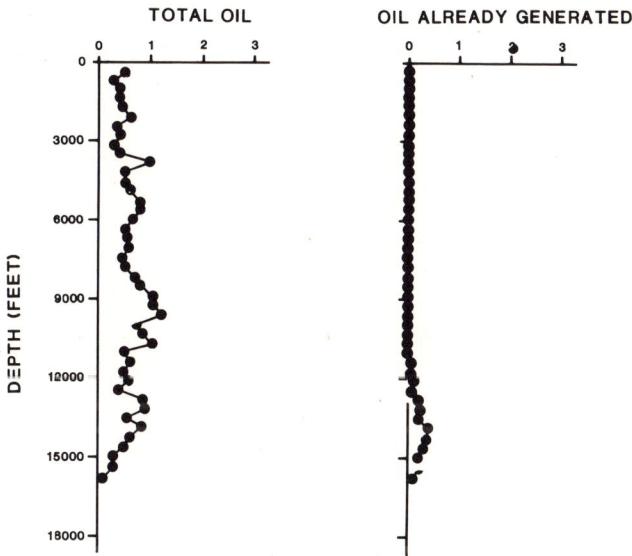

Figure 7.9. "Total Oil" and "Oil Already Generated" profiles for the C.O.S.T. #1 Well, offshore Texas Gulf Coast.

value of 0.5 (40 million barrels of bitumen generated per cubic mile of rock). It seems unlikely, therefore, that the penetrated section could be an excellent source for oil, regardless of its thermal history.

It would therefore be concluded that no excellent source rocks exist in this area, and that searching for deeper (and more mature) extensions of these rocks is unlikely to change this picture. At this point it would be wise to attempt to calculate the total volume of thermally mature rock in the region. This can be difficult, because often little is known about the most deeply buried sedimentary rocks in a basin. Nevertheless, if some assumptions and extrapolations are made, this calculation can be carried out. The total volume of bitumen which could have been generated in this region can then be estimated by multiplying the volume of mature rock (in cubic miles) by the yield per cubic mile (about 40 million barrels in this case). (The estimated total volume of bitumen must be reduced to compensate for inefficiency in expulsion, migration, and accumulation.)

The Tertiary rocks found in the Texas–Louisiana Gulf Coast oil province are very much like those penetrated in the C.O.S.T. #1 Well. No excellent source rocks have ever been identified, yet billions of barrels of oil have been produced there. As was noted in Chapter 5, the Gulf Coast is characterized by growth faults and excellent source rock-reservoir relationships, both of which promote efficient accumulation. The postulated high efficiency of migration in Gulf Coast sediments would more than compensate for the modest amounts of bitumen generated. Source rock studies must therefore be adjusted to take migrational efficiency into account in making any intelligent exploration decisions about the Gulf Coast.

6.3 Wilmington Field, Los Angeles Basin

A third example is taken from the Wilmington Field in the Los Angeles Basin. The samples whose analysis is represented in Figure 7.10 come from a single formation; the Puente Formation of Late Miocene age.

Quantity, as measured by TOC, is very high; the richest samples contain approximately 10% organic carbon. Quality was measured in two ways; both are shown in Figure 7.10. Immature H/C ratios of the kerogens range from 1.1 to 1.45 and indicate a very highly oil-prone kerogen. Pyrolysis yield, determined by a technique similar to that employed in the Rock-Eval, gives values of between 15 and 50 percent pyrolyzable organic carbon.

Figure 7.11 shows the "Total Oil" and "Oil Already Generated" profiles for this well. Because of the low degree of thermal maturity ($R_o < 0.6$ for even the deepest samples), "Oil Already Generated" is essentially negligible. "Total Oil," however, is very impressive for the samples from a depth of about 8000–9000 feet, and average to well above average for the shallower ones. Note that two different "Total Oil" profiles are shown. In the samples from around 6000 and 9500 feet, the discrepancies between the two analytical techniques are quite important. The profile based on H/C ratios is generally more optimistic than the one based on pyrolysis yields. In other words, the pyrolysis yields of these samples are much lower than would be predicted on the basis of the measured H/C ratios.

Which data are to be believed when an inconsistency like the present one emerges? This is a sticky question which arises all too often in source-rock studies. There are at least three possible explanations for this inconsistency:

- Sample contamination. It was noted earlier in this chapter that contamination is often a problem, especially in bitumen analyses. Kerogen contamination is less common, and may be due to the addition of such foreign material as walnut hulls or rubber during drilling, or the presence of solidified bitumen that can be mistaken for kerogen. Discrepant samples should be rechecked visually for signs of contamination.
- Analytical error. This is an ever-present possibility. Error may be random or **systematic, and its presence may be very hard to prove. Constant recalibration of** instruments against standards will help to minimize these problems.
- Poor interpretation. Much of our understanding of factors which influence petroleum generation is still rudimentary. Correlations of results from the various analytical techniques with quality and maturity factors may be erroneous. In the present case, perhaps an inappropriate correlation between pyrolysis yield and the quality factor was used. Or perhaps the H/C ratio–quality factor conversion scale is incorrect.

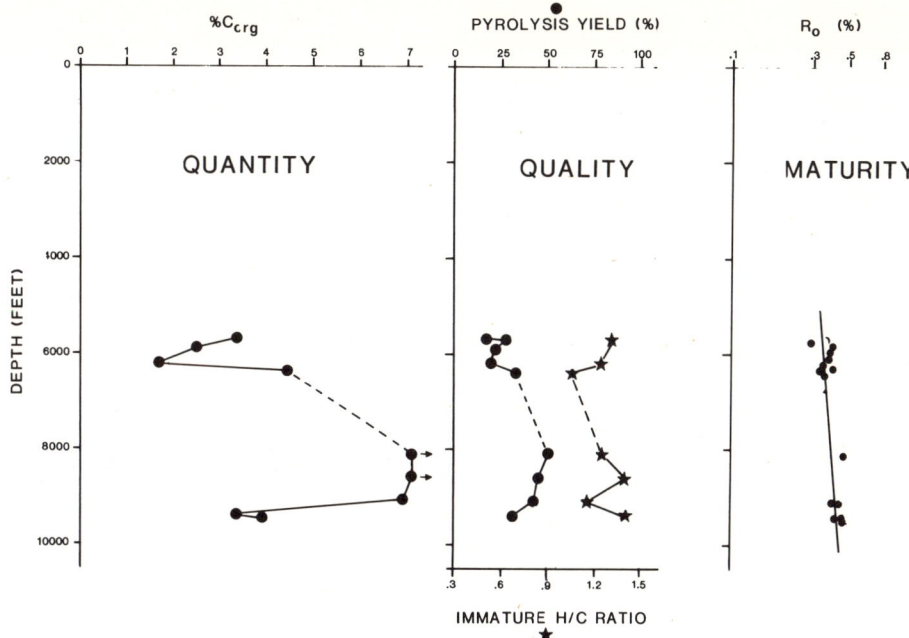

Figure 7.10. Kerogen quantity, quality, and maturity data for a Los Angeles Basin well. All samples analyzed are from the Puente Formation (Miocene).

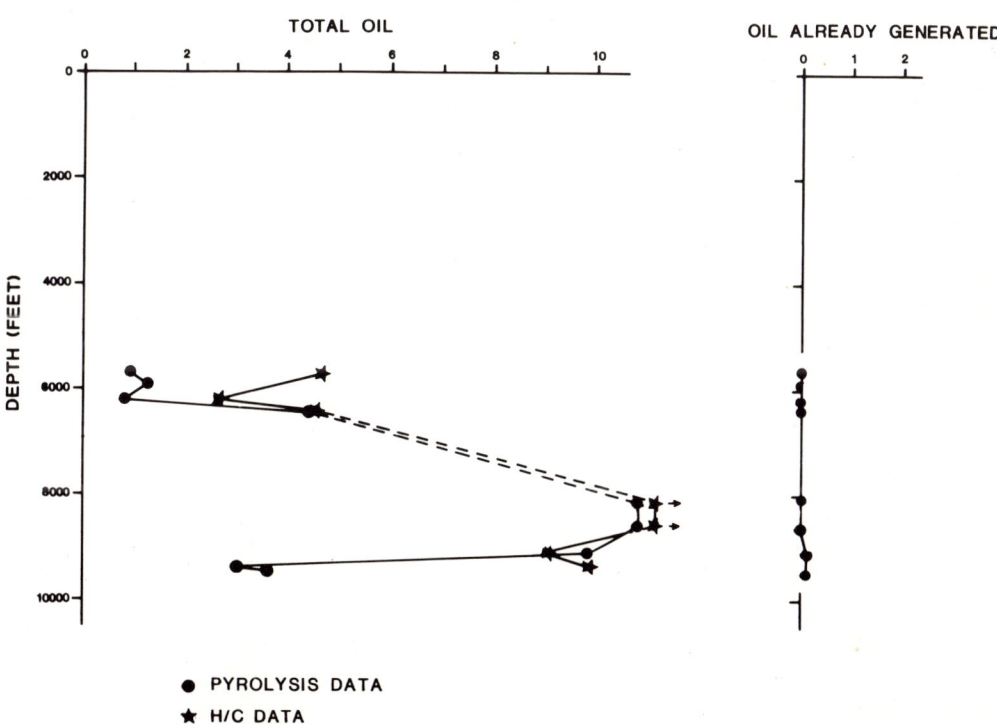

Figure 7.11. "Total Oil" and "Oil Already Generated" profiles for a Los Angeles Basin well, Puente Formation (Miocene).

The difficulties with data interpretation in the present example make one very important point: *it is always better to use several analytical techniques than to rely on a single one.* Multiple parameters serve as an internal system of checks and balances. For example, it is highly desirable to carry out MOA–TAI studies in addition to vitrinite reflectance measurements and H/C ratios. The human eye is a very sensitive organ and is quite useful in anticipating and explaining anomalous data. Unfortunately, for reasons of economy (often, in reality, false economy) many analysts tend to rely on single analytical techniques.

Regardless of which techniques are relied upon, a quick look at Figure 7.11 suggests that these sediments, if buried deeper, would have been superb source rocks. A first suggestion, then, would be to look basinward for a lateral extension of these Miocene sediments that might be a good source of oil.

The L.A. Basin Miocene shales are often highly fractured and contain large amounts of producible oil. They are generally thought to be the source rocks for the oil that they contain. Even low-maturity sediments, such as the ones analyzed in the present example, often are saturated with oil. Did this oil have a deeper, hotter origin in a basinward extension of these same sediments, or does the oil represent to a large extent an accumulation of indigenous, mostly diagenetic, bitumen?

If the oil has come from a deeper origin, then it represents a more or less normal case of bitumen generation and migration, although migration efficiency may be very high because of fracturing. If, however, the oil was generated in the very sediments in which it is now reservoired, then the analysis based on source-rock considerations alone has failed. To properly explain this oil occurrence, one of two adjustments to the interpretative procedure would be necessary:

- Correct the predictions of the source-rock analysis for the highly efficient migration which must occur when the source rock is also the reservoir.
- Take into account, especially in very organic-rich sediments, the initial (diagenetic) bitumen (5–10% of TOC) as a possible migratable fluid. None of the source-rock analyses presented until now have taken diagenetic bitumen into consideration. A simple adjustment to the maturity factor curve (Figure 7.6), in which perhaps 5 to 10% migratable bitumen exists even before the onset of generation, would take care of this, and probably would be an overall improvement on the model.

6.4 SUMMARY

Numerical analysis of geochemical data can give a semiquantitative prediction of oil source capacity. This analytical method can be employed in the manner outlined below.

1. "Total Oil" and "Oil Already Generated" must be defined clearly so that no confusion between these parameters exists.
2. The "Total Oil" value for any sample can be calculated from the equation

$$\text{"Total Oil"} = (Q_1)(Q_2)$$

where Q_1 and Q_2 are scaled parameters for quantity and quality of the kerogen in the sample.

3. The "Oil Already Generated" value for any sample may be calculated from the equation

$$\text{"Oil Already Generated"} = (Q_1)(Q_2)(M)$$

where M is the scaled parameter representing the degree of thermal maturity of the kerogen.

4. Because the "Total Oil" value of an "average" source bed is defined as 1.0, it is simple to interpret the "Total Oil" and "Oil Already Generated" values for any sample.
5. This method is well suited for well profiling because visual presentation of the results in graphical form facilitates interpretation.
6. The expulsion efficiency parameter, E, has not been studied in detail and is poorly understood. Inclusion of a reliable expulsion efficiency factor in the equations for "Total Oil" and "Oil Already Generated" would strengthen this approach.

7. Evaluation of Gas Source Capacity

Natural gas can be of either biogenic or thermal origin (see Chapter 3). Microorganisms in sediments produce methane in vast quantities. It is meaningless to try to treat this process quantitatively for exploration purposes, however, because the great majority of the methane is quickly lost to the atmosphere. The small fraction which may be trapped and accumulates in reservoirs is, however, significant to an explorationist. Perhaps 20% of the world's gas reserves are thought to be of biogenic origin, and in some regions—for example, the immense Urengoy Field in Siberia—the proportion may be much higher. Nevertheless, because accumulation of biogenic gas is much more of a trapping problem than one of generation, biogenic gas will not be considered further at this point.

Thermal gas is mainly formed after bitumen generation is complete, because it is more difficult to break off methyl and ethyl groups to form methane and ethane than it is to cleave longer hydrocarbon chains or to break carbon-heteroatom bonds (see Chapter 4). Methane

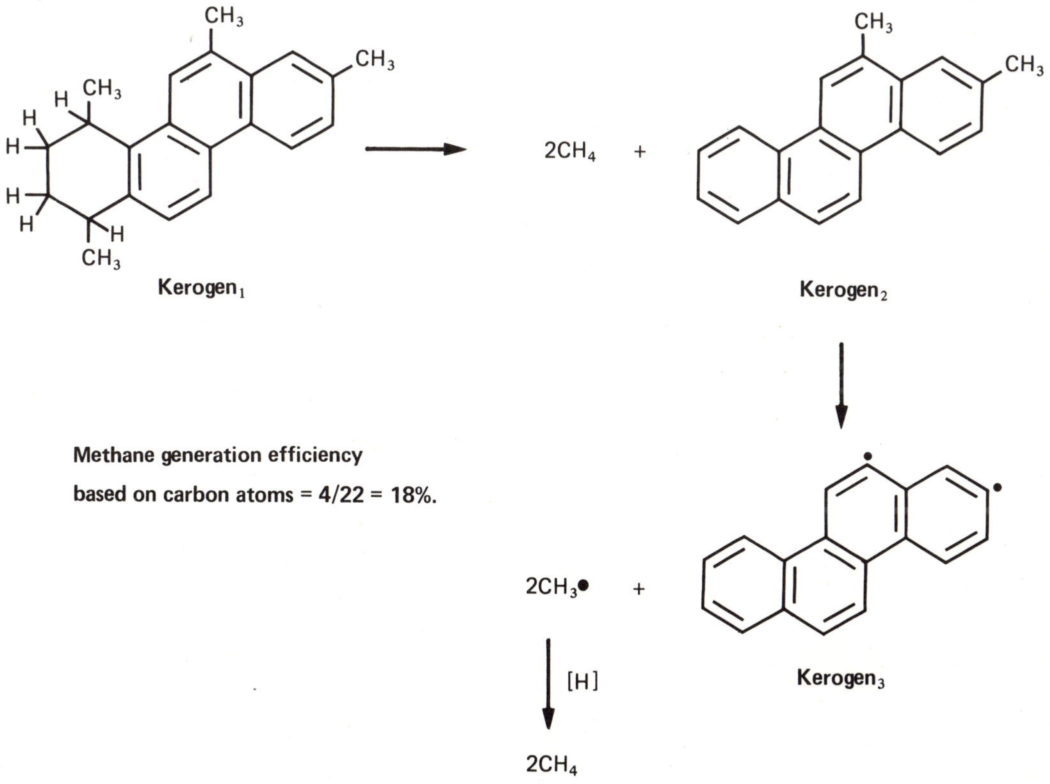

Methane generation efficiency
based on carbon atoms = 4/22 = 18%.

Figure 7.12. Formation of CH_4 from kerogens.

and the other small hydrocarbon molecules found in natural gas can be formed by thermal degradation of kerogen, bitumen, or petroleum. The total quantity of methane produced will depend on two factors: the H/C ratio of the organic precursor of the methane, and the degree to which this transformation has taken place.

Methane, CH_4, has the highest H/C ratio of any organic compound. Whenever a methane molecule is formed, some part of the residual precursor (kerogen, bitumen, or petroleum) molecule becomes depleted in hydrogen. It is this hydrogen depletion which limits the conversion of organic material to methane. Some illustrative examples are shown in Figure 7.12. In that figure, Kerogen$_1$ can lose two molecules of CH_4 relatively easily, because the increased stability brought on by aromatization promotes the demethylation reaction. After the CH_3–C bonds are broken, the methyl radical ($CH_3 \cdot$) formed abstracts a hydrogen atom (H·), forming a methane molecule and leaving a carbon–carbon double bond in

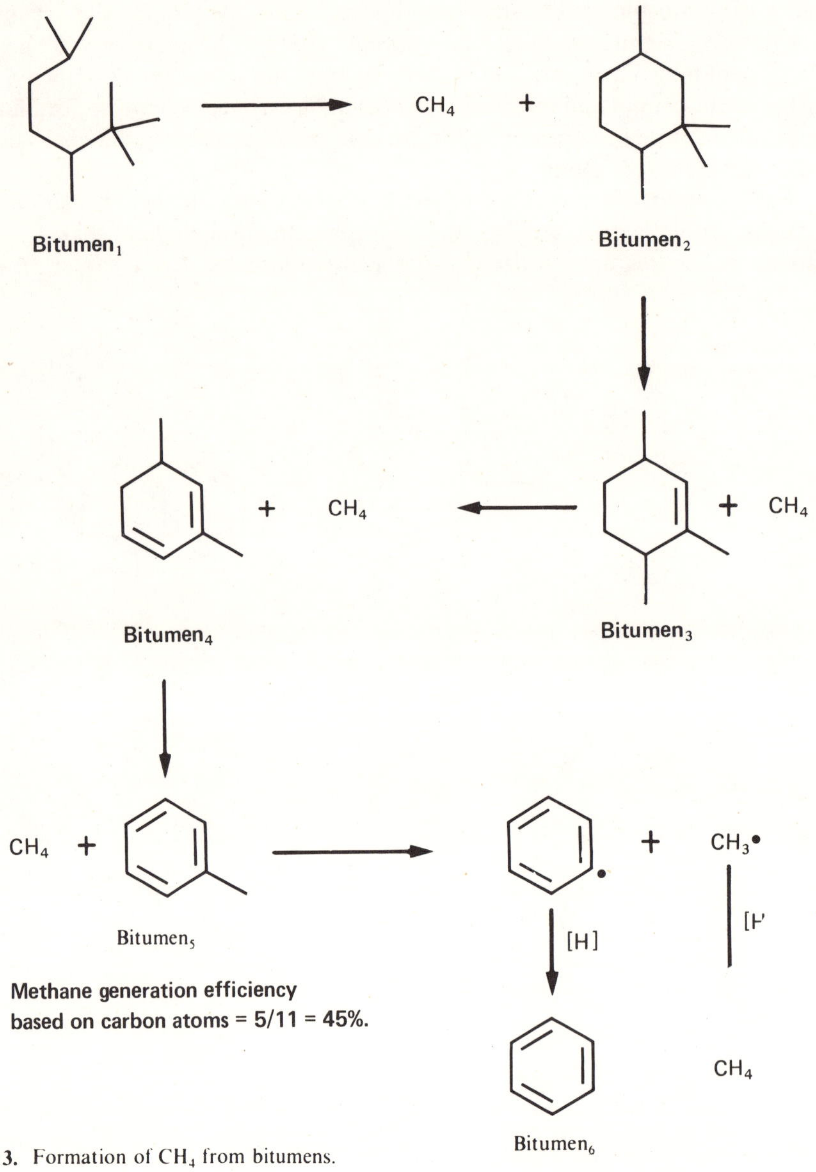

Methane generation efficiency based on carbon atoms = 5/11 = 45%.

Figure 7.13. Formation of CH_4 from bitumens.

the residual Kerogen$_2$. Kerogen$_2$ still has two methyl groups, but because the system is already fully aromatic, it cannot be further stabilized by losing another CH$_3\cdot$ and H$\cdot$. If the CH$_3\cdot$ does leave, it leaves behind an aromatic free radical. This free radical is somewhat stabilized by delocalization over the aromatic system (see Chapter 4). Despite delocalization, however, loss of methyl groups from Kerogen$_2$ is not as favored as it is from Kerogen$_1$, and it will occur later in the catagenetic sequence.

Because Kerogen$_1$ is hydrogen-poor, and Kerogen$_2$ is even more so, the methane yield per kerogen carbon atom is quite small (four methyl groups from twenty-two carbon atoms, or 18%). In general, the lower a kerogen's hydrogen content, the more aromatic it is, and the less methane it can ultimately generate.

Bitumen and petroleum can also yield methane during thermal cracking. Figure 7.13 shows a series of methane-forming reactions which could occur by thermal cracking of a branched hydrocarbon. The first loss of CH$_4$ results in cyclization, the next three gradually aromatize the ring, and the final one leads to an aromatic free radical. Because the odd electron is confined to a single benzene ring, the radical is less stable than the radical shown in Figure 7.12. The overall efficiency of methane generation in this example is 45%, much higher than that from the aromatic kerogen in Figure 7.12.

Although the above examples are somewhat artificial in that they represent the catagenetic behavior of single molecules rather than that of the complex mixtures present in kerogens and bitumens, they nevertheless are illustrative of the differences in methane-generative capacity of hydrogen-rich and hydrogen-poor precursors. The methane-generative efficiencies of 45% and 18% given in the examples are not experimentally established numbers for natural materials, but they may be fairly realistic estimates of the potential degrees of conversion of these materials to methane.

Kerogen						Bitumen					Kerogen + Bitumen	
Sample Depth	Q$_1$	Q$_2$	M	Gas Already Generated	Total Gas	Q$_1$	Q$_2$	M	Gas Already Generated	Total Gas	Gas Already Generated	Total Gas

Figure 7.14. Sample format for calculating gas-source capacity.

For years many people have insisted upon referring to certain types of kerogen as "oil-generative" and others as "gas-generative." This terminology conveys the incorrect impression that oil-generative kerogens cannot produce gas, and that gas-generative kerogens produce lots of gas. In fact, as was demonstrated above, a hydrogen-lean kerogen produces little gas. It is called gas-generative because the only thing it can yield is a little gas.

Oil-generative kerogens, on the other hand, produce large amounts of bitumen during catagenesis. This bitumen, if exposed to high enough temperatures for sufficient lengths of time, eventually will decompose as shown in Figure 7.13 to yield large amounts of gas. Thus, oil-generative kerogens are ultimately the best potential sources of natural gas also. This clarification is of great significance to exploration.

Gas-source capacity can be evaluated in a manner analogous to the way in which oil-source capacity is evaluated. It is necessary only to determine the quantity, quality, maturity, and expulsion and migration efficiency parameters of the source rock to be able to evaluate its gas generative capacity.

Because gas can be generated from both kerogen and bitumen, because the gas-source capacities of the two generally differ, and because bitumen is mobile while kerogen is not, these two gas-source materials must be analyzed individually. It would be best to use a format like that shown in Figure 7.14. Here again, the difficult expulsion and migration efficiency parameter has been omitted. Note that in calculating "Total Gas" based on contributions from both kerogen and bitumen, it is assumed that the bitumen will not migrate out of the rock, but instead will stay and be converted to methane. This is a very dangerous assumption, as is shown above in the discussion of migration. Calculations of gas-source capacity should be made with caution.

The quantity factor can be determined from the kerogen and bitumen contents. In this case, extraction of the bitumen would be necessary. Do not confuse TOC and BFOC here.

The quality factor could probably best be related either to pyrolysis data, especially for kerogens, or to elemental analyses (H/C) for both kerogen and bitumen. Electron spin resonance might also prove useful. All of these techniques require calibration before they can be applied to analyses of gas-source potential.

Maturity is best measured by either vitrinite reflectance or pyrolysis (Rock-Eval). TAI is inadequate in the gas-generation zone because the kerogen particles are quite dark. UV fluorescence is also of little value in this maturity range. The upper maturity limits for gas generation are not yet well defined.

At this time, analyses like that proposed above are rare or nonexistent. Nevertheless, in principle the question of gas-source capacity should be just as amenable to the approach outlined as is oil-source capacity.

8. Determining Growth and Depositional Environments

Determining the growth environment of the source organisms and the geochemical conditions in the depositional environment is also an important part of the regional aspect of exploration programs. If the growth and depositional environments can be related to kerogen quantity and quality, then a regional picture of source bed distribution as controlled by lateral facies changes will emerge. Knowledge of growth and depositional environments may also have significant implications for oil-source rock correlations, a topic which is discussed in Section 9.

Tissot et al. (1977) have discussed extensively the environmental significance of *n*-alkanes. A short summary of their conclusions appears in Chapter 4, Nevertheless, *n*-alkanes are only rough indicators of the total effects of growth and depositional environments, and should not be considered as conclusive proof of specific conditions without supporting evidence.

Two of the isoprenoids, phytane and pristane, have been used extensively as indicators of redox conditions in the depositional environment. A large excess of pristane over phytane appears to be good evidence for an oxidizing depositional environment (Brooks, 1970; Powell and McKirdy, 1975; Tissot and Welte, 1978, p. 387).

Steroids and triterpenoids are synthesized by specific organisms. Many steroids and triterpenoids are found only in certain kinds of organisms, so their presence is a diagnostic indicator of a contribution by those particular organisms. Among the most useful and least controversial generalizations are the following:

- Hopanes (a particular group of triterpanes) are synthesized mainly by prokaryotic organisms (bacteria and blue-green algae) and by terrestrial plants.
- In terrestrial material, the C_{29} steranes sitostane and stigmastane predominate over the C_{27} cholestane.

Other applications of sterane and triterpane data as environmental indicators must be regarded as somewhat speculative at present.

Carbon isotopes can in some cases be useful for determining the nature of organic source material. Land plants have been shown (Smith, 1975; Smith and Epstein, 1971) to have $\delta^{13}C$ values which fall into one of two groups. Most land plants metabolize CO_2 by the C_3 (Calvin cycle) biochemical pathway; these organisms have $\delta^{13}C$ values which fall in the range –24 to –34 $^o/_{oo}$ (versus PDB). A few land plants, especially those which grow under harsh conditions (swamps, deserts), metabolize CO_2 by the C_4 (dicarboxylic acid) pathway, and have much higher ^{13}C contents. $\delta^{13}C$ values for these plants are in the range –6 to –19 $^o/_{oo}$ versus PDB.

Marine plankton have $\delta^{13}C$ values which fall in the range for the relatively uncommon C_4 terrestrial plants, while freshwater plankton overlap with C_3 terrestrial plants. These generalizations are shown in Figure 7.15. Because most terrestrial plants belong to the C_3 group, and because most marine organic material is of planktonic origin (with $\delta^{13}C$ values like those for C_4 plants), it should in theory be possible to determine the origin of organic matter in rocks from its ^{13}C content.

In practice, however, application of carbon isotope data to environmental interpretations is rather complex, because many factors other than these general trends can influence the $\delta^{13}C$ values of fossil organic matter. The different classes of compounds (lipids, carbohydrates, etc.) have slightly different ^{13}C contents within a single organism. Selective removal of some classes of compounds may therefore alter the carbon isotope composition of the residue. Bacterial reworking could obliterate the carbon isotope pattern of the original organisms.

^{13}C content of any plant is dependent upon the ^{13}C content of the CO_2 it absorbs in photosynthesis. Different reservoirs of CO_2 (atmospheric versus CO_2 dissolved in water), different ^{13}C contents of atmospheric CO_2 over land and sea, and varying ^{13}C content of atmospheric CO_2 through geologic time (Welte et al., 1975) may make interpretation difficult. It is also likely that changes in the relative importance of the C_3 and C_4 pathways throughout the course of evolution have occurred.

Further care must always be taken in interpreting $\delta^{13}C$ values, because all incomplete chemical and physical processes result in some separation of the two isotopes. Microbial degradation or catagenesis can alter the $\delta^{13}C$ value of kerogen slightly. These effects, however, are usually not large enough to cause problems in evaluating the type of organic source material.

Occasionally, some samples of kerogen or oil have very unusual values of $\delta^{13}C$. Values in the range –18 to –21 $^o/_{oo}$ relative to PDB may indicate a hypersaline, highly reducing depositional environment (Galimov, 1973, ch. 3). Unambiguous correlations of carbon isotope ratios with specific depositional environments are rare, however.

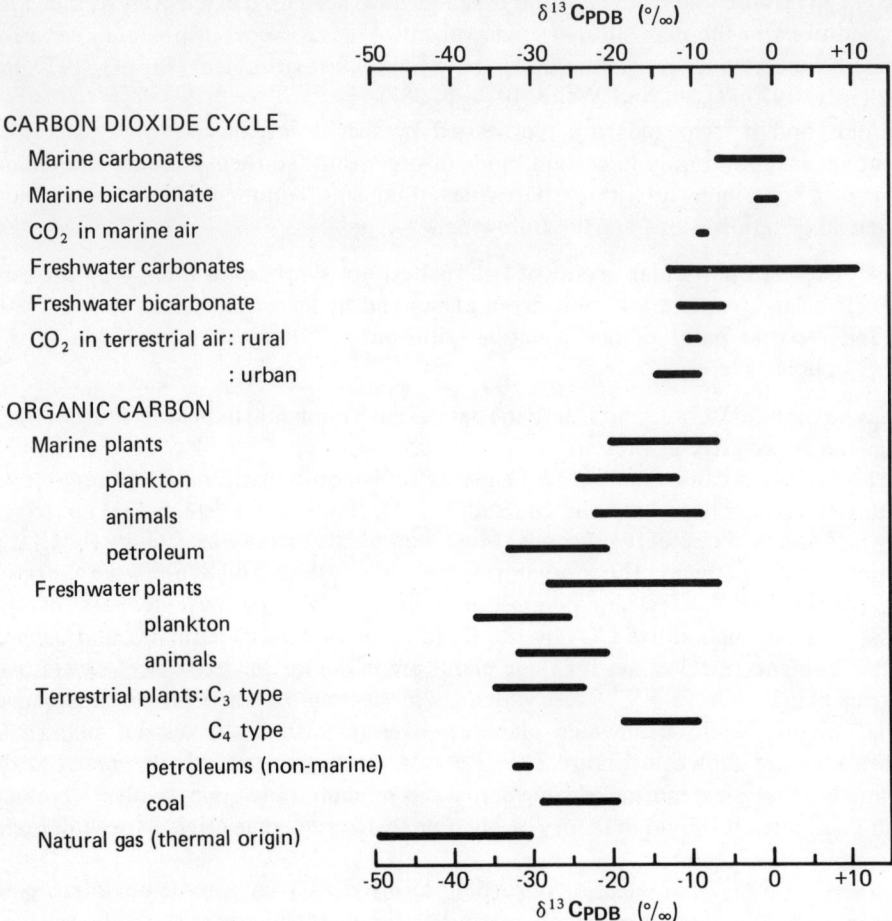

Figure 7.15. $\delta^{13}C$ ranges for various sources of organic and inorganic carbon versus PDB standard.

9. Correlations

9.1 INTRODUCTION

Oil–oil and oil–source rock correlations are of obvious importance to exploration, but surprisingly few conclusive correlations have ever been published. Oil–oil correlations are probably simpler, because one is comparing the same kind of organic material. Two oil samples having a common origin may differ substantially in chemical composition because of changes which have occurred during migration or storage in the reservoir rock. These changes can include loss of heavy, light, or polar components; biodegradation and water washing; and thermal disproportionation. In order to attempt oil–oil correlations, it is necessary to know how each of the above transformations will affect an oil's chemical properties. Milner et al. (1977) give a complete review of petroleum transformations in reservoirs.

Table 7.7 shows the effects of five types of alteration processes on crude oil composition. During migration, the heavier and more polar components are left behind, much as the polar compounds and asphaltenes are left behind during column chromatography (see Chapter 6). **Most of the migrationally induced changes in oil composition probably occur as the bitumen** makes its way out of the fine-grained source rock (primary migration). Secondary migration generally causes smaller changes.

Table 7.7. Effects of Alteration Processes on Crude Oil Composition

	Migration	Water Washing	Biodegradation	Gas Deasphalting	Thermal Maturation
API Gravity	↑	↓	↓	↑	↑
% Sulfur	↓	↑	↑	↓	↓
C_{15} + fraction (% of crude)	↓	↑	↑	↓	↓
Asphaltenes (% of crude)	↓	↑	↑	↓	↑ unless deasphalting occurs
Gasoline (C_4–C_7) fraction (% of crude)	↑	↓	↓	↑	↑
Paraffinicity	↑	↑	↓	↑	↑
Porphyrin content	↓	?	↑	↑	↓
n-Paraffins					
(a) % of crude	↑	↑ generally	↓	↑	↑
(b) maximum in distribution curve	↓ slightly	↑	↑ or no effect	No effect	↓
(c) CPI	No significant effect	No effect	↓ or no effect	No effect	↓
$\delta^{13}C$	↓	Depends on composition	↑	↓	↑ if gas is lost

Water washing and biodegradation often go together. Water washing can occur without biodegradation, but biodegradation will always be accompanied by at least some degree of water washing. During water washing, the more soluble components of petroleum are simply removed in solution. Light hydrocarbons, particularly aromatics, and the smaller polar molecules become depleted.

Microorganisms are very selective in which compounds they metabolize. Table 7.8 shows the classes of compounds that aerobic bacteria will consume, in order of preference. Anaerobic bacteria are not thought to be important in biodegradation of hydrocarbons. Compounds containing heteroatoms are often consumed relatively rapidly, and n-alkanes are generally severely depleted or totally absent in extensively degraded oils. Occasionally, the isoprenoid hydrocarbons are also noticeably depleted. Figure 7.16 shows gas chromatograms of a crude oil in the course of biodegradation simulated in a laboratory experiment.

Table 7.8. Hydrocarbons Arranged in Order of Increasing Resistance to Microbial Attack[a]

n-alkanes	C_{10}–C_{19}
n-alkenes	C_{12}–C_{18}
alkanes	C_2–C_4
branched alkanes	C_4–C_{12}
n-alkenes	C_{3-11}
branched alkenes	
aromatics	
cycloalkanes	

[a] After Perry and Cerniglia, 1973; republished with permission of LSU Press

Recently, Seifert and Moldowan (1979) have shown that biodegradation effects can be seen in the steranes and triterpanes. Certain of these compounds are more resistant than others, and these survivors can be used as fingerprints in biodegraded crudes.

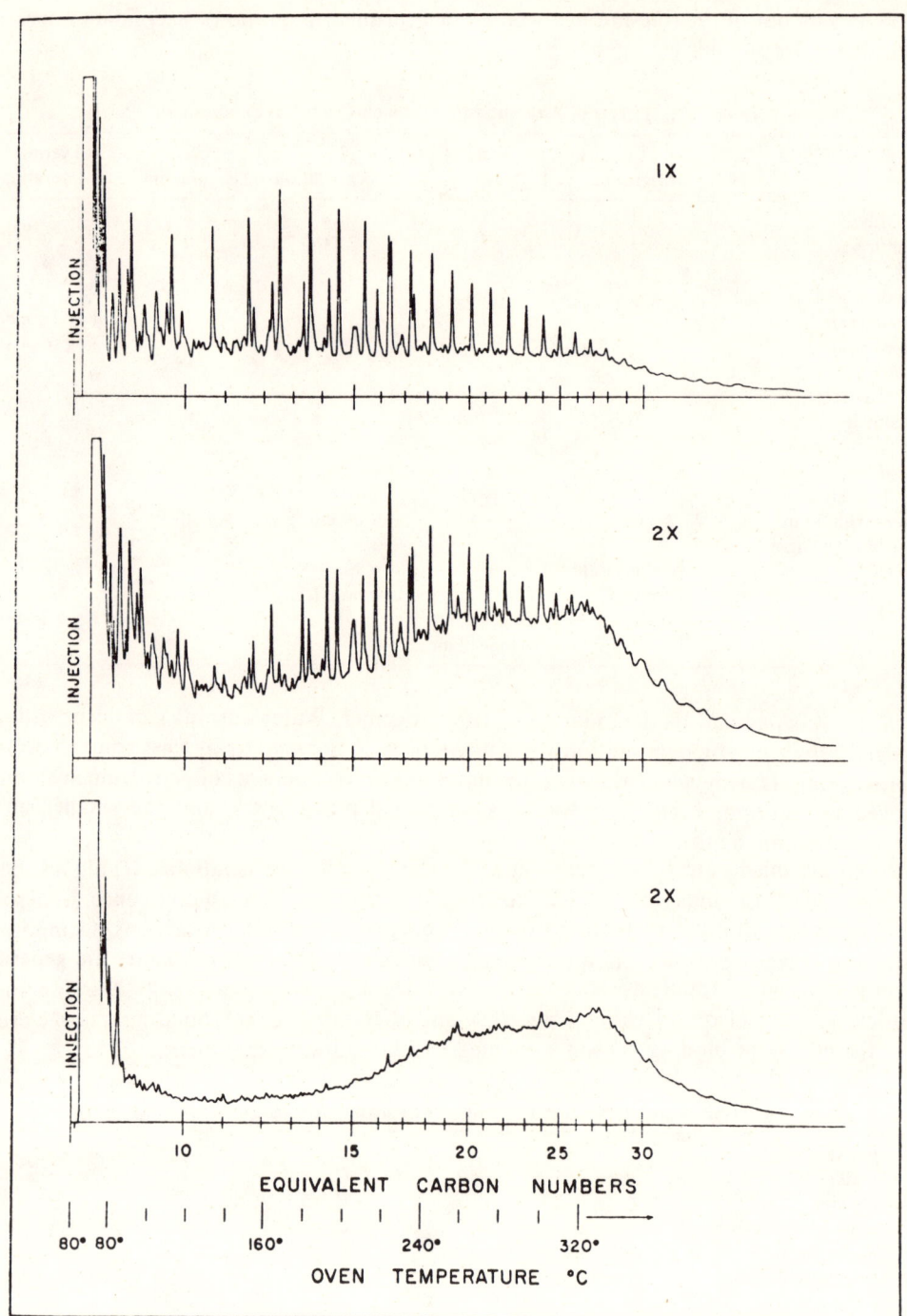

Figure 7.16. Gas chromatogram of crude oil at 0-time (top) and after 6 hours (center) and 26 hours (bottom) of incubation with a mixed population of microorganisms. Note changes in attenuation in top chromatogram. (From Mechalas et al., 1973; republished with permission of LSU Press)

Gas deasphalting results in the removal of the large aromatic molecules called asphaltenes. Because asphaltenes contain a significant proportion of the heteroatoms, deasphalting usually will lower the sulfur and nitrogen content of an oil. Thermal maturation results in disproportionation of middle-molecular-weight compounds to light hydrocarbons and asphaltenes, and is often accompanied by gas deasphalting.

Although the information in Table 7.7 is important, it does not tell the whole story. It is also necessary to know the magnitude of the changes which can be anticipated from the various alteration processes. Unfortunately, it is difficult to set rigid limits for most of these processes, because they range from mild to severe. Furthermore, the amount of change will depend somewhat upon the initial composition of the crude oil. In general, the best way to develop a feeling for the magnitude of change that one could reasonably expect is by doing many interpretations. Some guidelines are helpful, however.

1. API gravity may decrease fairly drastically during extensive water washing and biodegradation. Thermal maturation can cause very large changes in the opposite direction. Migration-related changes are less dramatic.
2. Sulfur content can sometimes be increased significantly during biodegradation.
3. Asphaltenes are very susceptible to loss during migration and to reservoir alteration (gas deasphalting and disproportionation).
4. The gasoline fraction **increases greatly during thermal maturation.**
5. Porphyrin content appears to be quite sensitive to migration. Nickel and vanadyl porphyrins seem to be affected equally, so the Ni/V ratio doesn't change much. Porphyrins are gradually destroyed by high temperatures, and hence by the refining process.
6. The content of n-paraffins may increase markedly with increasing maturity. Biodegradation can remove up to 100% of the n-paraffins. The maximum in the distribution curve often shifts very significantly toward lower carbon numbers with increasing maturity. Many immature crudes have maxima at C_{25} to C_{31}; at higher degrees of maturity the maximum could shift to C_{20}, C_{15}, or even lower. CPI gradually approaches 1.0 as oil maturity increases, but this change may be accompanied by drastic decrease in the amount of n-paraffins in the carbon range where the CPI was calculated. Two oils that have similar n-paraffin distribution curves but greatly differing CPI values probably are not from a common source.
7. $\delta^{13}C$ values change by only *at most* 1 or 2 $°/_{oo}$ during these transformations. This is due to two factors. If the system is closed, so that one compound is merely transformed into another compound, no ^{13}C or ^{12}C enters or leaves the system, so no net change in $\delta^{13}C$ can occur. Also, the different fractions (saturates, aromatics, polars, asphaltenes) seldom differ by more than 2 $°/_{oo}$ in ^{13}C content, so even the migrational loss of all the asphaltenes would not greatly affect the isotopic composition of the oil (Fuex, 1977).

9.2 OIL–OIL CORRELATIONS

The general approach taken in attempting a correlation between two oils is to measure several of the parameters listed in Table 7.7 and then compare the observed differences in the oils with those predicted by the table. It should first be noted that some properties of oils are very sensitive to genetic differences in oils; among these are carbon isotopes, sterane and triterpane contents, and Ni/V porphyrin ratios. Other properties, such as CPI and isoprenoid distributions, are somewhat dependent upon the origin of the oil, but may be influenced to a greater or lesser degree by such transformation processes as maturation, biodegradation, and migration. Still other properties (for example, API gravity and light hydrocarbon content) are mainly influenced by thermal and migrational transformations, and therefore are not very useful in oil–oil correlations. When carrying out oil–oil correlations, it is important to keep in mind which properties convey the maximum amount of genetic information.

For example, suppose that we suspect that Oil A is genetically related to Oil B, but that Oil B has probably undergone some biodegradation. To verify this hypothesis, we might analyze

Table 7.9. Analytical Data for Attempted Correlation of Oil A with Oil B

Parameter	Oil A	Oil B
API gravity	31.7°	26.6°
Ni/V porphyrins	1.1	1.2
n-paraffins (% of total crude)	18.6	12.2
n-paraffin distribution	see below	see below
δ^{13}C (topped oil) (°/$_{oo}$ vs PDB)	−27.3	−26.8

Oil A — n-Paraffin, carbon numbers 15 20 25 30

Oil B — n-Paraffin, carbon numbers 15 20 25 30

both oils for API gravity, porphyrin content, n-paraffin content and distribution, and ^{13}C/^{12}C. Suppose that we obtain the data shown in Table 7.9.

These data are all consistent with a common origin for oils A and B, but do not prove this conclusively. Oil B shows signs of slight biodegradation, having lost the lighter n-alkanes. Its API gravity has therefore decreased somewhat. Porphyrin ratios are about the same; we would not expect biodegradation to change those. δ^{13}C values are approximately the same.

When correlations are made, negative evidence is stronger than positive evidence. When data are incompatible with a certain hypothesis, we have proof that the hypothesis is not correct. But no amount of positive evidence can ever prove conclusively that two oils are related. Positive correlations are always based on circumstantial evidence.

In the example above there is no evidence which leads us to disbelieve our original hypothesis that the oils are related. On the other hand, the positive evidence is rather meagre, and by itself would not constitute proof beyond a reasonable doubt. Further analyses, particularly highly sensitive ones like comparisons of the steranes and triterpanes in the two oils, would greatly strengthen the case for a genetic relationship.

As a second example, consider two oils obtained several miles apart which are thought possibly to have originated from a common source rock. Oil D lies considerably farther from the proposed source basin than does Oil C. On the basis of the analytical data presented in Table 7.10, can it be determined if they did in fact come from a common source?

Many of the data support a common source. API gravities are close, and as might be expected, Oil D, which would have migrated further, is slightly lighter. Sulfur contents are comparable, but present a small problem because, contrary to the present example, sulfur content usually decreases in the course of migration. Oil D has a slightly lower proportion of the heavier (C_{15}+) components, a trend which is compatible with the migration hypothesis. Porphyrin ratios, like sulfur contents, are similar, but the trend is in the opposite direction to that which would be expected from migration effects. In any case, the difference is fairly small and could be within experimental error. Carbon isotopes, however, present a serious problem. The two oils differ by 3.3 °/$_{oo}$. No known migrational effects could be responsible for such a large change in δ^{13}C. We conclude, therefore, that Oils C and D do not have a common source, and that the observed similarities are fortuitous.

Another possibility is that Oil D is actually a mixture of oils from two sources. One could be the common source for Oil C, and the other a completely different source with a very different isotopic composition.

Williams (1974) has used a multivariate approach like that above to classify oils in the Williston Basin, and to correlate them with three possible source rocks. The more variables one

Table 7.10. Analytical data for attempted correlation of Oil C with Oil D

Parameter	Oil C	Oil D
API Gravity	32.6	36.7
% Sulfur	1.08	1.27
$C_{15}+$ (% of total crude)	37.5	34.8
Ni/V porphyrin ratio	0.65	0.81
$\delta^{13}C$ (°/$_{oo}$ vs PDB)	-26.3	-29.6

can measure, the more persuasive one's case is for a correlation. With large numbers of samples and analyses such as Williams used, some sort of computerized cluster analysis can be very helpful.

Koons et al. (1974) used four parameters to separate Lower Tuscaloosa oils into two groups. The analyses they carried out were for light hydrocarbons, $C_{15}+$ hydrocarbons, paraffin contents, and carbon isotope ratios. In the light hydrocarbon fraction, they used the ratio of cyclopentanes to n-paraffins, and in the heavy hydrocarbons, the ratio of C_{28} steranes to C_{27} steranes. Powell and McKirdy (1975) classified Australian and Papuan oils into three groups on the basis of wax content, correlation indices, pristane/phytane ratios, and geologic environment.

The parameters which one chooses to use when attempting a correlation or a grouping of oils will depend in part on the particular analyses that can readily be carried out, and in part on what seems to work. Powell and McKirdy (1975) found wax content and pristane/phytane ratios to be useful, because much of the source organic material was derived from land plants. In other geologic settings, these particular parameters might not be useful at all.

9.3 Oil–Source Rock Correlations

Oil–source rock correlations are somewhat more difficult, because one must compare an unmigrated bitumen with a petroleum. Compositional differences might be expected to be great, even if there were a genetic relationship. The approach one takes is the same as that used for oil–oil correlations, but some parameters are less useful. API gravity, for example, is not significant for a bitumen.

Williams (1974) used carbon isotope data, n-alkane distributions, and light hydrocarbon distributions in correlating his three Williston Basin oil types with three source rocks, and Welte et al. (1975a) used several parameters in correlating a world-wide suite of oils with their known source rocks. Mathews et al. (1970), on the other hand, placed most of their emphasis on hydrocarbon distributions in bitumen and petroleum in an attempt to identify the source rock for Moonie Field oils in Australia. It would be much better to examine several other independent parameters to lend additional support to their conclusions.

Stahl (1978) has recently proposed that by measuring the $^{13}C/^{12}C$ ratios of the saturate, aromatic, polar, and asphaltene fractions of crude oils, one can estimate the $^{13}C/^{12}C$ ratio of the source kerogen. This method will increase the utility of carbon isotope measurements in source rock–oil correlations.

Because bitumen and petroleum differ so dramatically in composition, it is very difficult to make definitive source rock–oil correlations. Welte et al. (1975a), in fact, compiled data on most of the certain and suspected correlations which existed world-wide through 1974. The geochemical properties most sensitive to genetic relationships, such as carbon isotope ratios (Stahl, 1978) and polycyclic hydrocarbons (Seifert and Moldowan, 1978) are becoming increasingly important in such correlations.

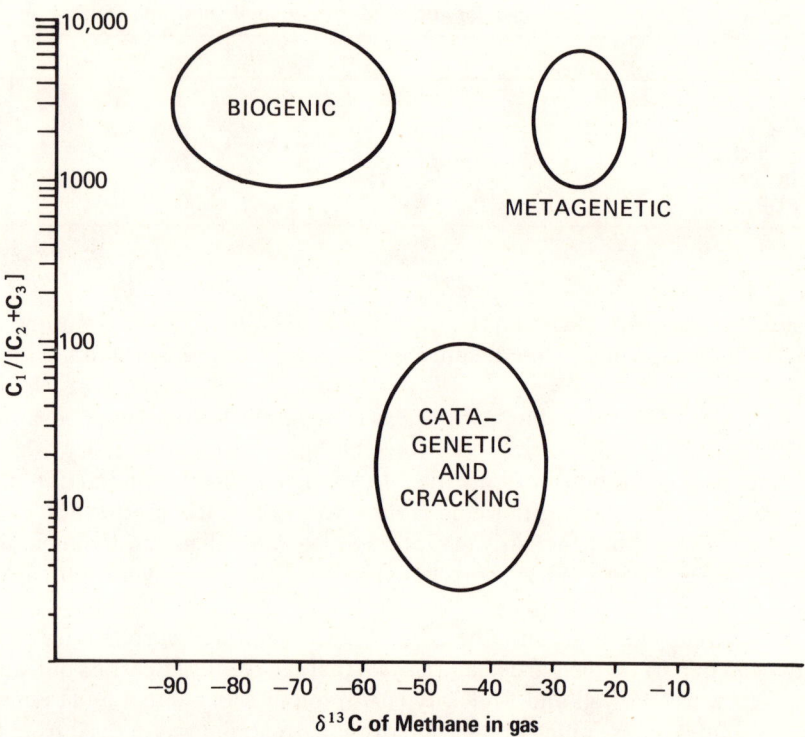

Figure 7.17. Classification of natural gases by plotting wetness against $\delta^{13}C$. Open areas represent mixtures of gases from more than one source.

9.4. Natural Gas

It is not possible to correlate one sample of natural gas with another with any degree of confidence because the mixtures are so simple and because they have so few distinguishing characteristics. It has, however, been possible to establish the origin of natural gases by measurements of two properties: wetness and ^{13}C content.

Carbon-13 measurements are particularly important because they can distinguish biogenic gas from thermally generated gases. Biogenic methane is highly depleted in ^{13}C, having ^{13}C values from about -55 to $-90\ ^o/_{oo}$ relative to PDB. Thermally generated methane, in contrast, has ^{13}C values below $-55\ ^o/_{oo}$.

Wet gases are formed only during catagenesis and by cracking of petroleum or bitumen. Both biogenic and metagenetic (very high temperature) gases are dry, consisting almost exclusively of methane. It is therefore useful to plot the ^{13}C value of the methane in natural gas against the wetness in order to identify the origin of the gas (Bernard et al., 1976; Sackett, 1977; Fuex, 1977; Stahl, 1974, 1975). Figure 7.17 shows such a plot and the regions where gases of single origins are found.

Biogenic and metagenetic gases are both dry, but have very different carbon isotopic compositions (Figure 7.17). Gases formed by catagenesis of kerogen or by cracking of petroleum are wet. These three categories are the end-member classes. Gases of other compositions are considered to be the result of mixing of gases from more than one source. The wetness parameter is particularly sensitive, because only a small amount of wet gas need be present to greatly alter the composition of a biogenic or metagenetic dry gas.

8 TIME AND TEMPERATURE AS FACTORS IN OIL GENERATION*

1. Introduction

It is well documented that oil generation is promoted by high subsurface temperatures. This observation is in keeping with the predictions of chemical reaction-rate theory. The Arrhenius Equation (Eq. 8.1) gives the exact dependence of the rate constant (k) on the activation energy (E_a) and the temperature (T).

$$k = Ae^{-E_a/RT} \tag{8.1}$$

The pre-exponential factor A is a constant, the exact value of which depends upon the particular reaction under consideration, and R is the universal gas constant.

Several workers have calculated activation energies for the process of oil generation. To do this, they made an Arrhenius plot for several different basins in which oil had been generated. (See any physical chemistry text for information on Arrhenius plots.) The values obtained for E_a are in the range of 11,000–14,000 calories per mole (Tissot, 1969; Connan, 1974). Welte (1972), however, pointed out that activation energies in this range are far lower than one would expect for the breaking of carbon-carbon or carbon-oxygen bonds (40,000–60,000 cal/mole) and many authors (e.g., Connan, 1974; Shimoyama and Johns, 1971, 1972) have interpreted the low values of E_a as proof of the importance of mineral catalysis in oil generation.

In one respect this appeared to be a reasonable hypothesis, because catalysts lower activation energies by providing alternative, lower-energy pathways. One problem with the catalyst idea, however, is that no catalysts are known which could lower the activation energies for individual kerogen decomposition reactions to 11,000–14,000 cal/mol. Most catalytic effects are far less dramatic.

In 1975, Jüntgen and Klein published a paper which presented a much more plausible explanation. They pointed out that oil generation involves many parallel reactions, and that the overall rate of oil generation should depend upon the sum of the rates of all the parallel chemical reactions which produce bitumen molecules. When the individual reactions are summed, but the process is treated mathematically as though it were a single reaction, the calculated activation energy turns out to be much lower than the activation energy of any of the individual reactions. As Jüntgen and Klein (1975) showed, it is not really a true activation

---

*Much of this discussion is taken from Waples (1980).

energy at all, but rather a mathematical construct. This is why the "pseudo-activation energy" for oil generation has the chemically impossible value of 11,000–14,000 cal/mol. The interested reader is referred to the articles by Jüntgen and Klein (1975) and Snowdon (1979) for further discussion.

Early efforts to take both time and temperature into account in studying the process of oil generation were only partially successful because of the mathematical difficulties inherent in allowing both time and temperature to vary independently (Connan, 1974; Waples, 1976). In 1971, however, N.V. Lopatin published a paper which described a simple method by which the effects of both time and temperature could be taken into account in calculating the thermal maturity of organic material in sediments. He developed a "Time-Temperature Index" of maturity (TTI) to quantify his method.

Lopatin's original work was greeted with some enthusiasm and much criticism. Some of the problems which surfaced could be attributed to the poor quality of the data with which Lopatin originally calibrated his model. In spite of these technical details, however, Lopatin's basic idea has merit, for it is capable of predicting the thermal conditions under which hydrocarbons can be generated and preserved.

2. Construction of the Geologic Model

Implementation of Lopatin's method begins with a reconstruction of the depositional and tectonic history of the geologic section of interest. This is best accomplished by plotting depth of burial against geologic age, as in the hypothetical example shown in Figure 8.1. It should always be remembered that such reconstructions are *not* geologic cross-sections. In the example in Figure 8.1, a Lower Cretaceous sediment was deposited 125 m.y. B.P. at the

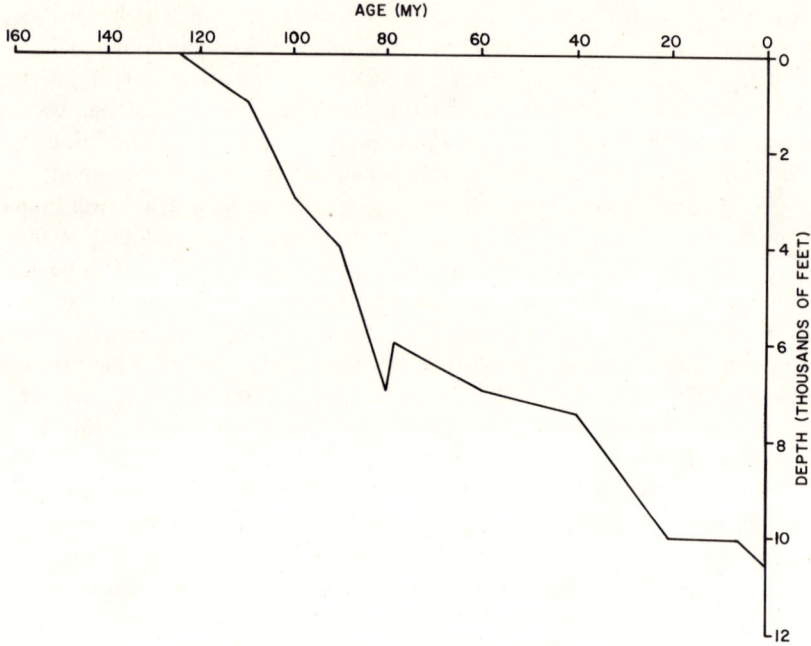

Figure 8.1. Depositional and tectonic history of a Lower Cretaceous sediment.

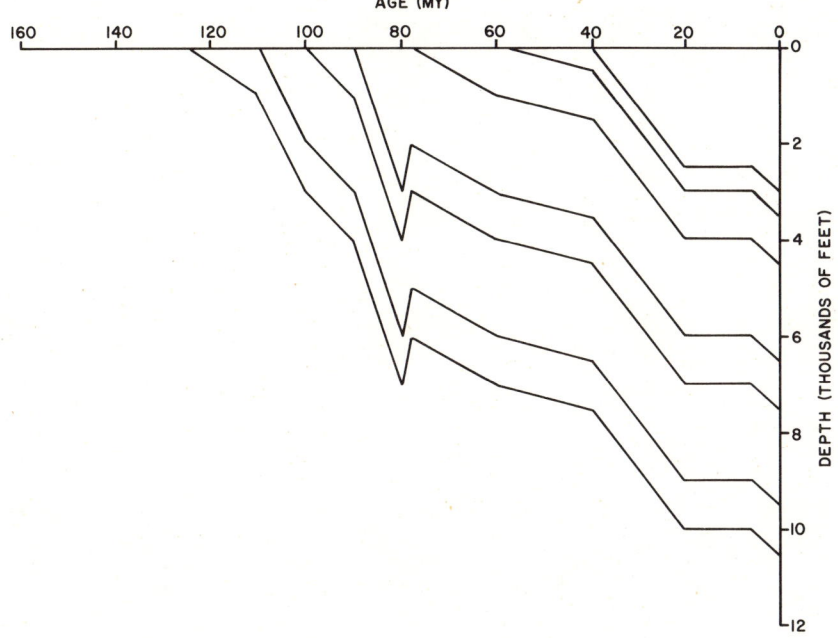

Figure 8.2. Depositional and tectonic history of several sedimentary horizons.

sediment surface (depth = 0). Since its deposition, the sediment has had the time–depth history shown by the solid line in the figure. Its history consisted of continual deposition at varying rates until 80 m.y. B.P., at which time there occurred a brief (2 m.y.) uplift, during which the sediment was raised from a depth of 7,000 feet to a depth of 6,000 feet. Uplift was followed by renewed subsidence until a depositional hiatus was reached at 20 m.y. B.P. The hiatus persisted until 6 m.y. B.P., at which time subsidence commenced again. The sediment is now (time = 0 m.y. B.P.) at a depth of 10,500 feet. The line in Figure 8.1 thus traces the depth–time relation for the sediment. Any shallower sediments, such as those shown in Figure 8.2, will have depth–time lines subparallel to the first line, commencing with their own deposition. A set of these

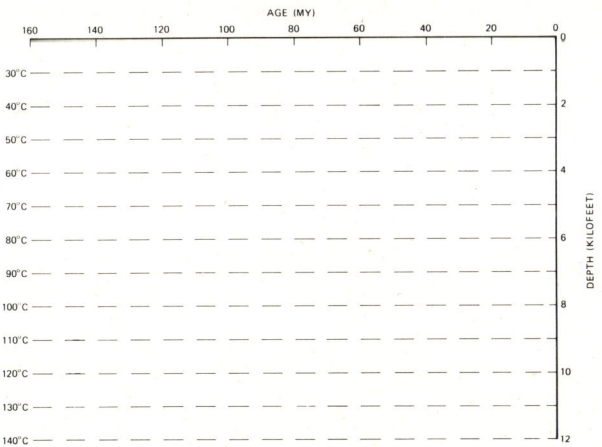

Figure 8.3. Simple subsurface temperature grid.

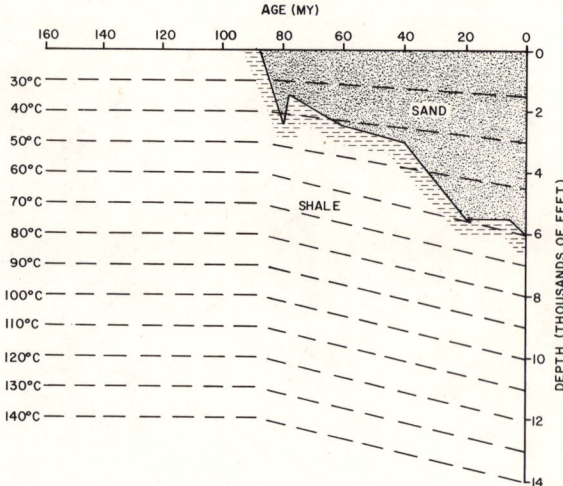

Figure 8.4. Complex subsurface temperature grid.

lines, such as those in Figure 8.2, forms Lopatin's geologic reconstruction. Except in certain special situations, which are considered later in this chapter, the depth-time lines for the various horizons will always be parallel.

The geologic reconstruction is based on the best information available to the geologist. In some cases, particularly where deposition has been continuous, it is easy to make the reconstructions with a high level of confidence. In the cases of sediments which have complex histories, the reconstruction may represent only the best guess possible.

The second aspect of the geologic model is the temperature grid. The subsurface temperature must be specified for every depth throughout the geologic past. The simplest way to do this is to compute the present-day geothermal gradient and assume that both the gradient and the surface temperature have been constant throughout the time interval covered by the reconstruction. An example of such a temperature reconstruction is shown in Figure 8.3.

Figure 8.4 shows a more complicated situation, in which there is a break in the present-day geothermal gradient. The upper part of the section, which is mainly sand, has a low gradient, while the lower, shaly part has a high gradient. If it is assumed that the past geothermal gradients were also related to lithology, the geothermal gradient prior to 88 m.y. B.P. was high

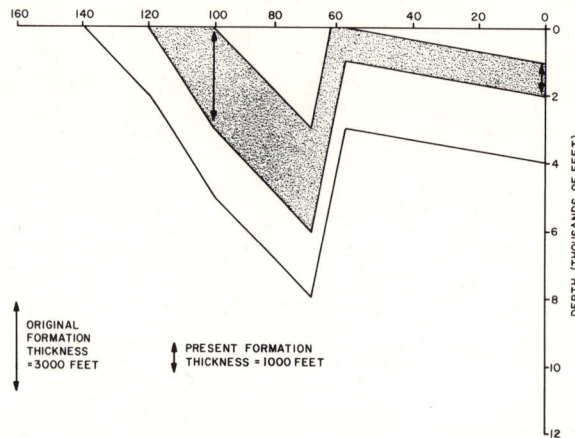

Figure 8.5. Illustration of section thinning by erosion.

for the entire section, because only shales were present. The low gradient came into existence after 88 m.y. B.P., when decomposition of sand began. The isotherms (dashed lines) in Figure 8.4 thus represent the subsurface temperature as a function of geologic time.

There is no theoretical limit to the complexity which can be introduced into the temperature history of a section. In most cases, however, the data necessary for a highly sophisticated temperature reconstruction are simply not available.

It should be noted that Lopatin's method can be applied to any geologic model, regardless of the model's crudeness or complexity. A well-thought-out, detailed reconstruction will obviously yield more reliable results than one which is largely based on guesswork. These limitations should be borne in mind in any subsequent interpretation of Lopatin data. In some cases, even a very crude or approximate model may be able to answer important questions about hydrocarbon generation or preservation.

3. Special Cases

Although in many cases the geologic models can be constructed in a straightforward manner, there are some situations in which caution is advisable, or where special techniques are necessary. When uplift and erosion occur, some section is lost. Thus, while the horizon lines remain parallel after such an event, the distance between them will be reduced, as illustrated in Figure 8.5.

Another problem can arise when the section under examination is cut by a fault. In these cases, the sections above and below the fault may have had different thermal histories. It is necessary therefore to make two different geologic reconstructions for the two different sides of the fault, and combine them in order to obtain the complete reconstruction for the section. An example of a combined reconstruction for a normally faulted section is shown in Figure 8.6. The horizons are parallel within each of the two parts of the reconstruction, but the horizons in the footwall need not be parallel to those in the hanging wall. The volume of rock represented by the shaded area is common to both sections, but is not shown explicitly in this profile. In thrust faulting, where a section is repeated, it will usually be necessary to put the upper and lower reconstructions on separate charts to avoid confusion. The principle, however, is the same as when a section is missing.

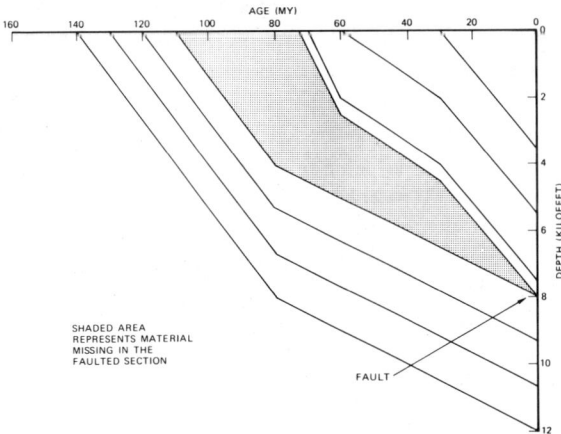

Figure 8.6. Loss of section by faulting.

Table 8.1. Temperature Factors for Different Temperature Intervals

Temp. Interval (°C)	n	Temperature Factor, γ
30–40	–7	$2^{-7} = 0.0078$
40–50	–6	$2^{-6} = 0.0156$
50–60	–5	$2^{-5} = 0.0313$
60–70	–4	$2^{-4} = 0.0625$
70–80	–3	$2^{-3} = 0.125$
80–90	–2	$2^{-2} = 0.25$
90–100	–1	$2^{-1} = 0.5$
100–110	0	$2^{0} = 1$
110–120	1	$2^{1} = 2$
120–130	2	$2^{2} = 4$
130–140	3	$2^{3} = 8$
140–150	4	$2^{4} = 16$
150–160	5	$2^{5} = 32$
....	m	2^{m}

Igneous intrusive events can, in principle, be taken into account by adjusting the geothermal gradient during the period of heating. Unfortunately, the temperatures reached in the sediments and the durations of the intrusive events are usually poorly known, so it is difficult to make an accurate time–temperature reconstruction.

4. Theory of Lopatin's Method

Lopatin and many others believe that two factors, time and temperature, are important in oil generation and destruction. These two factors are interchangeable; a high temperature acting for a short time can have the same effect on maturation as a low temperature acting over a longer time. Lopatin assumed that the dependence of maturity on time is linear; that is, doubling the cooking time at a constant temperature doubles the maturity.

Chemical reaction-rate theory predicts that the rate of a reaction that occurs at about 70 °C with an activation energy of 17,000 calories per mole will double with every 10 °C increase in temperature. These temperature and activation-energy values are similar to those calculated empirically by Connan (1974) for the oil-generative process. This correlation suggests that Lopatin's assumption is valid. Recent data (Waples, 1980) confirm its validity, although other workers (Neruchev and Parparova, 1972; Golitsyn, 1973) have challenged it.

Lopatin divided the temperature profile into 10 °C intervals and drew the isotherms in the manner shown in Figures 8.3 and 8.4. He chose the 100–110 °C interval as the base interval, and assigned to it an index value of $n = 0$. The other intervals were assigned the index values shown in Table 8.1. Lopatin then defined a γ factor, which reflects the exponential dependence of maturity on temperature. He assumed that the rate of maturation increased by a factor of 2 for every 10 °C increase in reaction temperature. Thus within any 10 °C temperature interval $T_i - T_{i+1}$, the temperature factor γ is equal to 2^n, where $n = (T_i - 100)/10$ (see Table 8.1).

For his time factor, Lopatin used the length of time (in m.y.) that the sediment spent in each temperature interval. The maturity added in any temperature interval i is given by

$$\text{Maturity}_i = 2^{n_i} \Delta T \tag{8.2}$$

where ΔT_i is the length of time spent by the sediment in the ith temperature interval.

Because maturation effects on the organic material are additive, the total maturity (or TTI) of a given sediment is equal to the sum of the maturities acquired in each interval. Thus

$$\text{TTI} = \sum_{n_{\min}}^{n_{\max}} 2^n \Delta T_n \tag{8.3}$$

where $n_{\max}$ and $n_{\min}$ are the n-values of the highest and lowest temperature intervals encountered. If Lopatin's idea is correct, the TTI value should correlate with data obtained from other methods of evaluating the thermal maturity of organic material.

5. Calculation of TTI

The principles involved in calculating TTI values have been explained above; a specific example of such a calculation is given here. Figure 8.7 shows a geologic model with three sediment horizons (A, B, and C) and a moderately complex temperature grid. The present-day TTI values are calculated in the manner shown in Table 8.2. In all cases ΔT for a particular interval is merely the age at which the sediment enters that interval, minus the age at which it enters the next interval.

The same procedure can be used to calculate the TTI value at any time in the past. Suppose that we are interested in the TTI value of horizon A 60 m.y. ago (Point P in Figure 8.7). The calculations are carried out in a manner analogous to that above, but stop 60 m.y. B.P. instead of at the present time. The calculated TTI value for point P in Figure 8.7 is 5.9.

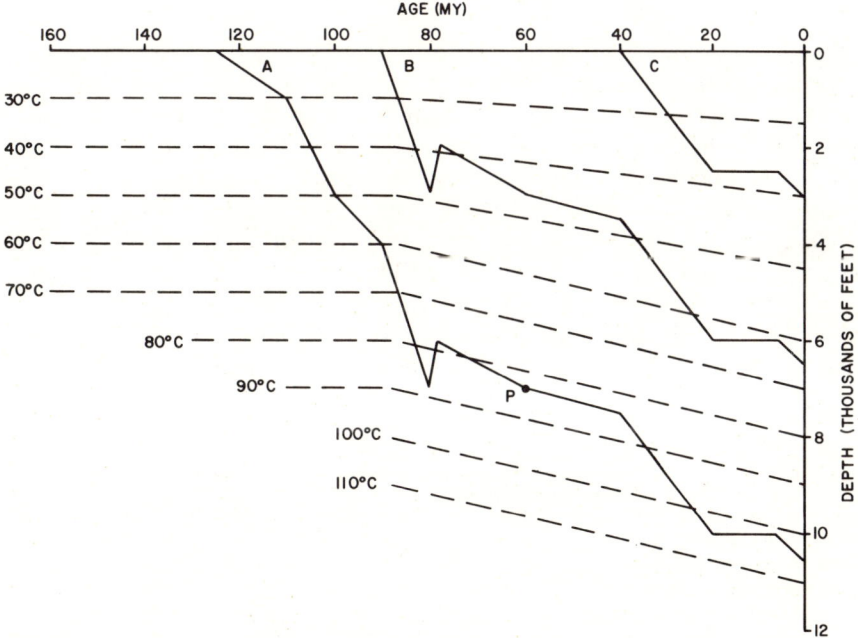

Figure 8.7. Geologic reconstruction for three horizons.

Table 8.2. Calculation of TTI Values

Temp. Interval (°C)	γ	Time (m.y.)	Interval TTI	Total TTI
Horizon A				
20–30	2^{-8}	15	0.06	0.06
30–40	2^{-7}	5	0.04	0.10
40–50	2^{-6}	5	0.08	0.18
50–60	2^{-5}	10	0.31	0.49
60–70	2^{-4}	3.5	0.22	0.71
70–80	2^{-3}	(3.5+6.5)	1.25	1.96
80–90	2^{-2}	(4.5+37.5)	10.5	12.5
90–100	2^{-1}	10.5	5.3	17.8
100–110	1	24	24	41.8
Horizon B				
20–30	2^{-8}	3.5	0.01	0.01
30–40	2^{-7}	(3.5+2.5)	0.05	0.06
40–50	2^{-6}	(5 + 38)	0.67	0.73
50–60	2^{-5}	12.5	0.39	1.12
60–70	2^{-4}	24.5	1.53	2.65
Horizon C				
20–30	2^{-8}	10.5	0.17	0.17
30–40	2^{-7}	29.5	0.22	0.39

6. Interpretation of TTI Values

Lopatin (1971) originally proposed that definite TTI values correspond to the different stages in the process of oil generation. The specific values that he chose, however, appear to be incorrect because of errors in his original geologic reconstruction. A new scale of TTI values has therefore been constructed by comparing the measured vitrinite reflectance (R_o) and TAI with TTI values calculated from 31 geologic models from a wide variety of geologic environments and ages throughout the world.

Table 8.3 shows R_o, TAI, and TTI values for several important stages of oil generation and preservation. These limits effectively define the TTI range in which oil generation occurs (15–160), the highest TTI values at which oils of 40° and 50° API will be preserved (500 and 1,000, respectively), and the highest TTI values at which wet gas can be preserved (1,500). Dry gas is produced in the Union of California #1-33 Bruner, Beckham Co., Oklahoma, from a horizon

Table 8.3. Correlation of TTI with Several Important Stages of Oil Generation and Preservation

Stage	TTI	R_o	TAI
Onset of oil generation	15	0.65	2.65
Peak oil generation	75	1.00	2.9
End of oil generation	160	1.30	3.2
Upper TTI limit for occurrence of oil with API gravity 40°	500	1.75	3.6
Upper TTI limit for occurrence of oil with API gravity 50°	1,000	2.0	3.7
Upper TTI limit for occurrence of wet gas	1,500	2.2	3.75
Last known occurrence of dry gas	65,000	—	—
Liquid Sulfur in Lone Star Baden #1 (Below dry gas limit)	972,000	5.0	4.0

Table 8.4. Time Stratigraphy for C.O.S.T. #1 Well, Texas Gulf Coast

Age (my)	Depth (ft)
0.5	1100
1.8	2400
3.7	3450
5.0	3800
10.0	10,100
11.0	10,500
14.0	14,000
15.0	15,700

that has a TTI of about 65,000, but it has not yet been established that this is the maximum possible TTI at which methane is stable.

7. Correlation of TTI with Other Geochemical Data

Calculated TTI values were correlated with measured data on several accepted or proposed maturity parameters: TAI, vitrinite reflectance, Bitumen/C_{org} ratios, CPI, H/C ratios, percent expandable clays, and API gravity. Correlations of TTI with each of these parameters led to the same conclusion: TTI is a valid measure of thermal maturity of organic material. A more complete discussion of these results is found in a recent article (Waples, 1980).

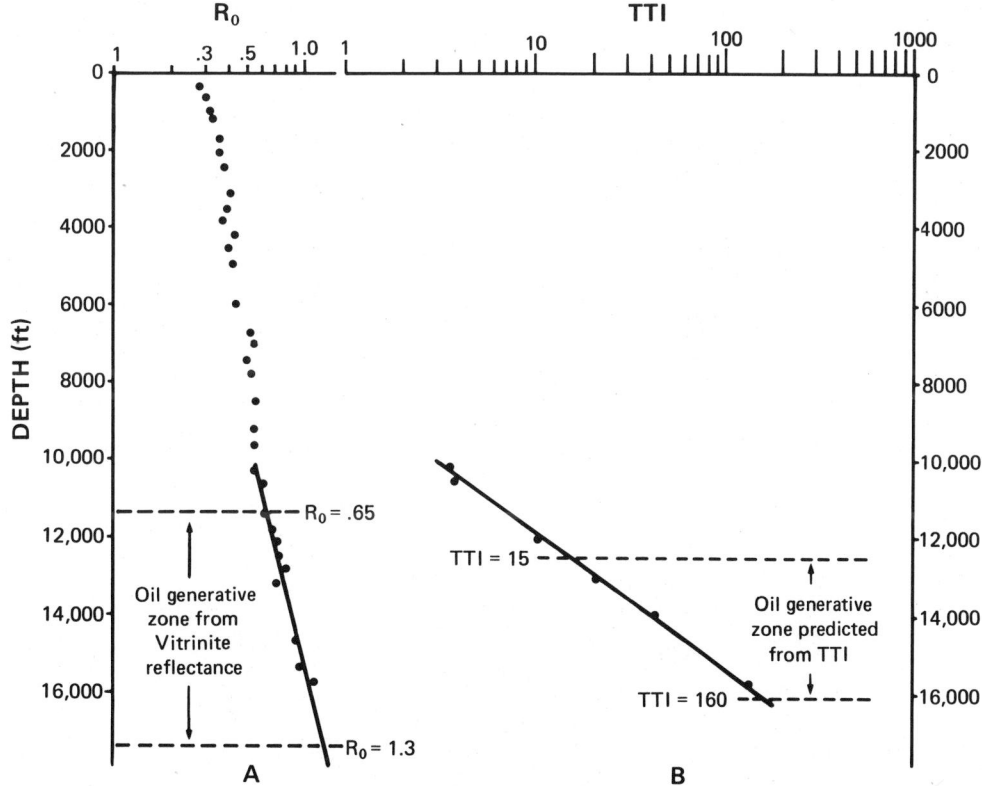

Figure 8.8. Comparison of oil-generative zones for C.O.S.T. #1 well, Texas Gulf Coast. A, from measured vitrinite reflectance values; B, from calculated TTI values.

As an example of the satisfactory correspondence between measured geochemical maturation parameters and calculated TTI values, let us look at an example cited in Chapter 7, the C.O.S.T. #1 Well, from the Texas Gulf Coast. The time stratigraphy used in the Lopatin reconstruction is shown in Table 8.4. The geothermal gradient was assumed constant at 1.75 °F/100 feet, which is the present-day gradient in the bottom half of the well. The surface temperature was taken as 64 °F. Figure 8.8 compares the vitrinite reflectance values and TTI values as functions of depth.

There is a generally good agreement between the vitrinite reflectance and TTI values. The measured reflectance values reach the onset of oil generation (R_o = 0.65%) at a slightly shallower depth than do the calculated TTI values. This discrepancy could reflect our use of too low a geothermal gradient in the upper part of the hole.

There is also some apparent disagreement about the end of oil generation. A TTI of 160 is reached at 16,100 feet, whereas a simple extrapolation of the vitrinite reflectance data suggests that full maturity is not reached until 17,400 feet. Extrapolation of this particular reflectance curve is difficult, however, because of the lack of data in the mature region. With these uncertainties in mind, the agreement between measured and predicted values is not bad.

8. Application of TTI Data to Exploration

TTI values obtained by application of Lopatin's method can be useful in several ways for oil exploration. If one is concerned with how deep preserved accumulations of oil, wet gas, or dry gas can be expected, one need only calculate the present-day TTI values of the suspected reservoirs and determine the TTI regime into which they fall. For example, suppose

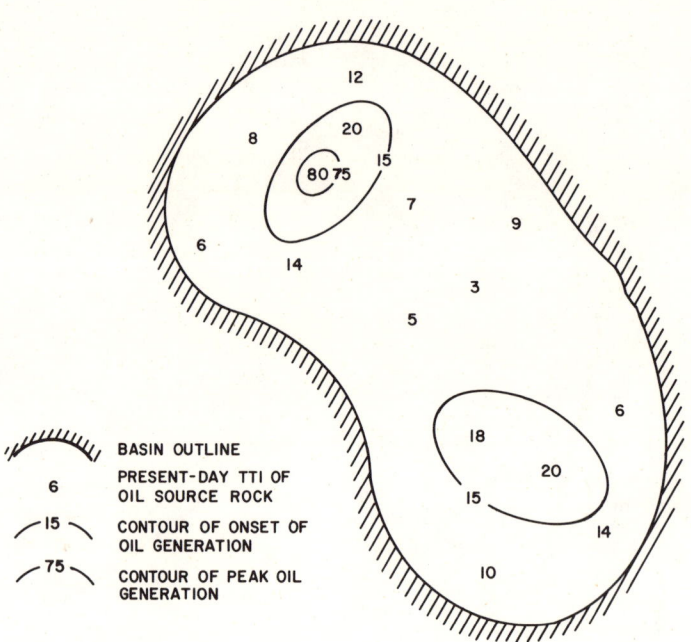

Figure 8.9. Present-day TTI values of an organic-rich shale in a hypothetical basin.

it is expected that a certain reservoir rock will be encountered at 12,000 feet in a proposed well. Assuming the existence of a suitable oil-source rock, can oil or gas be expected in the well, and if oil, of what gravity?

Suppose that a TTI of 1,200 is calculated for the reservoir formation. This means that the reservoir has a higher TTI value than that at which a 50° oil can be preserved (1,000; see Table 8.2). The TTI calibrations therefore predict that the reservoir lies beyond the oil deadline, and would likely contain only wet or dry gas. As stated previously, the confidence level of this interpretation would depend upon the quality of the geologic model. This interpretation also depends upon the assumption that oil migrates upward.

A second way in which TTI values can assist in oil exploration is in answering the question of whether or not the thermal maturity necessary for bitumen generation has occured in a region, For example, suppose an organic-rich shale has been found in a basin, and it is important to know whether this shale has reached thermal maturity. By making time–depth reconstructions for several points in the basin, present-day TTI values can be calculated for the shale at these points, as shown in the hypothetical example in Figure 8.9. By contouring the TTI values, one can get an idea of the areal extent of rich shale which has entered the generative window. In the example in Figure 8.9, the generative area (within the TTI = 15 contours) represents only a small portion of the total basin; hence only a small fraction of the rich shale could have begun to generate oil. Thus, the exploration risk in prospects adjacent to this basin would be considerably higher than if the whole basin had already reached thermal maturity.

A third application of TTI data in exploration is in answering questions about the timing of generation. Figure 8.10 shows a geologic model in which TTI values of 15 and 160 have been located on each of several horizons. If one contours iso-TTI values (Figure 8.10), one finds two lines which delimit the oil-generative window for the entire section throughout the geologic past. The shaded region in Figure 8.10 indicates the generative window. Suppose that one particular formation, indicated as "Oil Source Rock" in Figure 8.10, is the only known oil-source rock for this region. The time in the geologic past at which the OSR generated oil can be

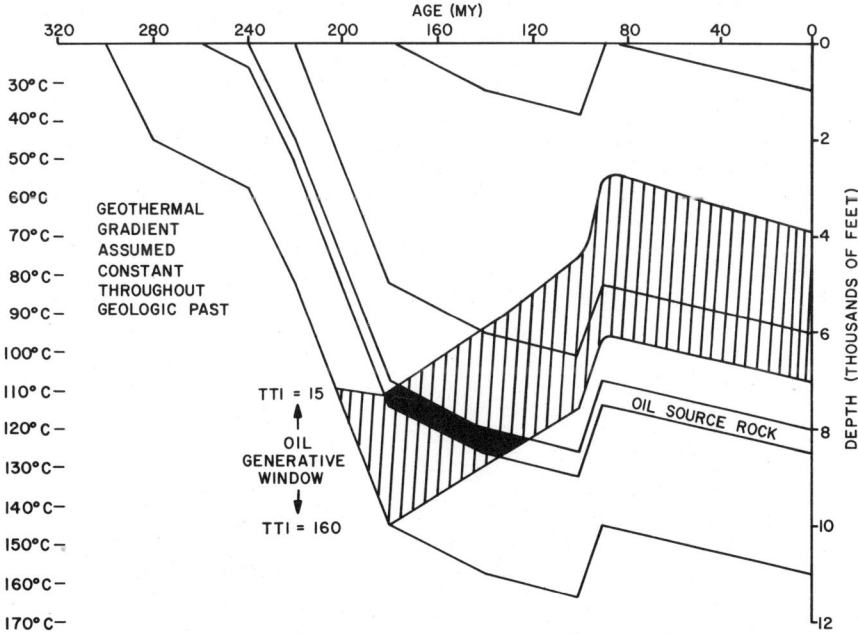

Figure 8.10. Iso-maturity lines on a geologic reconstruction.

determined by inspection of Figure 8.10, where the time–depth conditions for oil generation are shown in black. The OSR entered the generative window 181 m.y. B.P. and ceased generating oil 120 m.y. B.P. Because the time span during which oil generation occured is now known (from 181 to 120 m.y. B.P.), one can begin to answer important questions about the timing of oil generation and trap formation. Suppose that in this case the only structural traps in the region were created during the uplift lasting from 100 to 90 m.y. B.P. Because trap formation occurred at least 20 m.y. after the end of oil generation, the probability is low that this oil could have been captured by these local traps. It is more likely that by the time these traps were formed, the oil had already migrated out of the region, because there existed no barrier to its movement.

Finally, Lopatin's method can also be used to derive information about paleotemperatures and tectonics. For example, if the calculated TTI values from a given well correlate poorly with measured maturity data, there is a good possibility that either the paleotemperatures were different from the temperatures assumed in the model, or that assessment of uplift and erosion during and after tectonic events was not done correctly. The paleotemperature and tectonic interpretations can then be reevaluated and a revised reconstruction checked against the measured data. In this way different proposed histories can be tested, leading to an increased understanding of otherwise obscure geologic events in the region.

This list of potential applications of Lopatin's method is doubtless incomplete, because the method is very versatile. Zieglar and Spotts (1978) have applied Lopatin's method successfully in a study of California basins, and MacMillan (1980) has used the technique in the Rocky Mountains. Other creative geologists will undoubtedly discover new ways in which they can use TTI data to answer specific questions of importance in their own particular exploration areas.

9 PRACTICE PROBLEMS

The following problems are arranged approximately in order of increasing difficulty and sophistication. It is recommended that you attempt them in the order given. Keep in mind that the problems will be of most benefit to you if you attempt them before reading the solutions.

1. Source-Rock Evaluation

Problem 1

Combine the data from the Blue Well given in Table 9.1 to give a coherent picture of thermal maturity in the section drilled. Explain how you resolved any apparent discrepancies.

Table 9.1. Thermal maturity data for the Blue Well

Depth (ft)	TAI[a]	R_O[b]	Bit/BFOC[a]
1000	2.0	—	0.05
1200	2.0	—	0.07
1500	2.0	—	0.02
2000	2.0	—	0.10
2300	2–2.5	—	0.08
2600	2.0	—	0.09
3000	2.3	—	0.06
3200	2.3	—	0.17
3400	2.0	—	0.25
3700	2–2.5	0.42	0.49
4000	2.2	0.49	1.03
4200	2.5	0.46	0.86
4800	2.5	0.55	0.21
5000	2–2.5	0.60	0.03
5200	2.6	0.51	0.07
5400	2.5	0.59	0.09
5700	2.5	0.63	0.11
6000	2.6	0.60	0.12

[a] TAI = Thermal Alteration Index
[b] R_O = Vitrinite reflectance
[c] Bit/BFOC = Bitumen/Bitumen-free organic carbon

Solution. Each of the three parameters is related to maturity. As was discussed in Chapter 4, the bitumen/kerogen ratio increases in the oil-generative zone. If the Bitumen/BFOC ratios given in Table 9.1 were plotted, it would be noted that the values increase dramatically between 3200 and 4800 feet. This increase in bitumen content could mean that oil generation is occurring within this zone. It could also indicate the presence of contamination from drilling fluid additives, or could be the result of migration of bitumen into these strata from some other source.

One clue in the data suggests that the bitumen in the 3200–4800 foot zone is not indigenous: the Bitumen/BFOC ratios are much higher than would be expected from oil generation alone. During oil generation, these ratios usually go from 0.05–0.10 to 0.10–0.20. That is, the change is much less than that in the Blue Well samples. It is very difficult to attribute bitumen/kerogen ratios in excess of 0.4 to catagenesis.

The kerogen data support the conclusion that no significant oil generation has occured in this section. TAI values reach only the lowest threshold of oil generation (2.6) in the deepest samples. Vitrinite reflectance (R_o) values are consistent with the TAI values. Neither indicates sufficient maturity for oil generation in the 3200–4800 foot interval. We conclude therefore that the entire section investigated is thermally immature.

If one wanted to determine whether the high bitumen contents in the 3200–4800 foot interval are the result of migration or contamination, gas-chromatographic analysis of the hydrocarbon fraction of the bitumen could be useful. Furthermore, the composition of the bitumen might give clues to its origin. Contamination often comes from diesel fuel or other refined products added to the drilling fluid. Compared to crude oils, these materials have high proportions of hydrocarbons, particularly those with fewer than 20 carbon atoms. The heteroatom contents of refined materials are also very low.

Problem 2

Perform a source-rock analysis on the Mauve Well, data for which are given in Table 9.2.

Table 9.2. Source-rock data for Mauve Well

Depth (m)	Type of Sample	%C$_{org}$	Atomic H/C	TAI[a]	% Alginite + Exinite
1000	Sidewall	0.6	1.07	2.0	75
1200	Cores	0.8	1.22	2–2.5	80
1500	↓	0.5	1.05	2–2.5	80
1750		0.3	0.65	2–2.5	75
2000		1.3	0.77	2.2	80
2300		0.7	0.81	2.6	90
2700	↓	1.6	1.33	2.5	85
3000	Core	2.5	1.27	2.5	75
3500	Cuttings	0.5	1.15	2.6	70
3600		1.2	0.98	2.7	50
3800		1.0	0.86	2.9	45
4000		0.7	0.75	3.0	60
4500		1.5	0.72	3.1	45
4600		1.7	0.66	3.2	40
4800		2.1	0.41	3.7	?
5000	↓	2.2	0.38	3.8	?

[a] TAI = Thermal Alteration Index.

Solution. Data are available on quantity (%C_{org}), quality (H/C and %Alginite + Exinite), and maturity (TAI), so "Total Oil" can be plotted against "Oil Already Generated." Two independent quality measurements have been made, and both should be utilized and examined for possible discrepancies. To use the H/C data, however, one must first convert the measured, present-day H/C ratios to the ones that the kerogens had when they were thermally immature. This can be done easily by plotting H/C versus TAI, as shown in Figure 9.1 (derived from Figure 7.1), and then tracing the H/C ratio back to its immature value. The calculated immature H/C ratios are listed in Table 9.3.

Table 9.3. Scaled Quality Data for Mauve Well Samples

Depth (m)	Measured H/C	Immature H/C	Quality Factor (from H/C)		Quality Factor (from macerals)
1000	1.07	1.07	1.05	*	1.5
1200	1.22	1.22	1.50		1.6
1500	1.05	1.05	1.00	*	1.6
1750	0.65	0.65	0.17	*	1.5
2000	0.77	0.77	0.35	*	1.6
2300	0.81	0.81	0.43	*	1.8
2700	1.33	1.35	1.85		1.7
3000	1.27	1.30	1.70		1.5
3500	1.15	1.20	1.35		1.4
3600	0.98	1.05	0.90		1.0
3800	0.86	1.05	0.90		0.9
4000	0.75	0.90	0.60	*	1.2
4500	0.72	0.90	0.60	*	0.9
4600	0.66	0.90	0.60		0.8
4800	0.41	?	?		?
5000	0.38	?	?		?

*Indicates discrepancy between quality factors calculated from H/C and from maceral analysis.

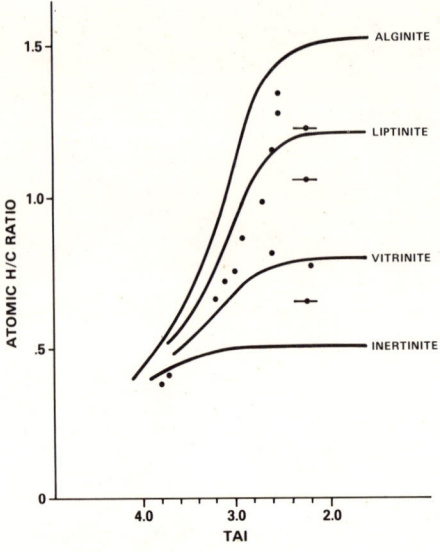

Figure 9.1. H/C versus TAI for Mauve Well samples.

Both the immature H/C ratios and the maceral analysis data need to be scaled to calculate "Total Oil." To do this, refer to Figure 7.2 and Table 7.5, respectively. The scaled quality factors are given for each parameter in Table 9.3.

It is apparent that there are serious discrepancies between the H/C and maceral analysis results for several of the samples. The samples at 1000, 1500, 1750, 2000, 2300, 4000, and 4500 meters all show differences in the quality factors calculated from the two types of data. In each case, the H/C ratio gives the lower quality factor, so some systematic error is likely. Without more knowledge, however, it is impossible to pinpoint the error. The prudent interpreter might now ask that some of the H/C ratios be rerun, to check for analytical error, and would certainly request that the slides made for maceral analysis be reviewed. If these attempts produced no resolution of the problem, the interpreter might then decide to try a third technique, such as pyrolysis. The most important point being made here is that these discrepancies must be taken seriously by the interpreter, and not be overlooked or swept under the rug. It may be necessary occasionally to offer two alternative interpretations without choosing between them. Let us take this last approach to this problem.

The rest of the section shows a good correspondence between the two parameters, except for the two deepest samples. These two kerogens are highly mature and quite black. In fact, no maceral analysis was possible here, and the H/C ratios are not helpful because the maceral types cannot be ascertained from such low H/C values. One can say little, therefore, about the oil-source history of the section below 4600 meters.

"Total Oil" and "Oil Already Generated" profiles are plotted in Figure 9.2. Most of the discrepancies among the different quality factors turn out to be unimportant, because source-rock potential is not good for most of the section. The only sample where the discrepancy is significant is that from 2000 meters. "Total Oil" values are generally unexciting, although the section between 2000 and 3500 meters shows fairly good potential. More samples between 3000 and 3500 meters should be obtained to define better the zone of high "Total Oil" values.

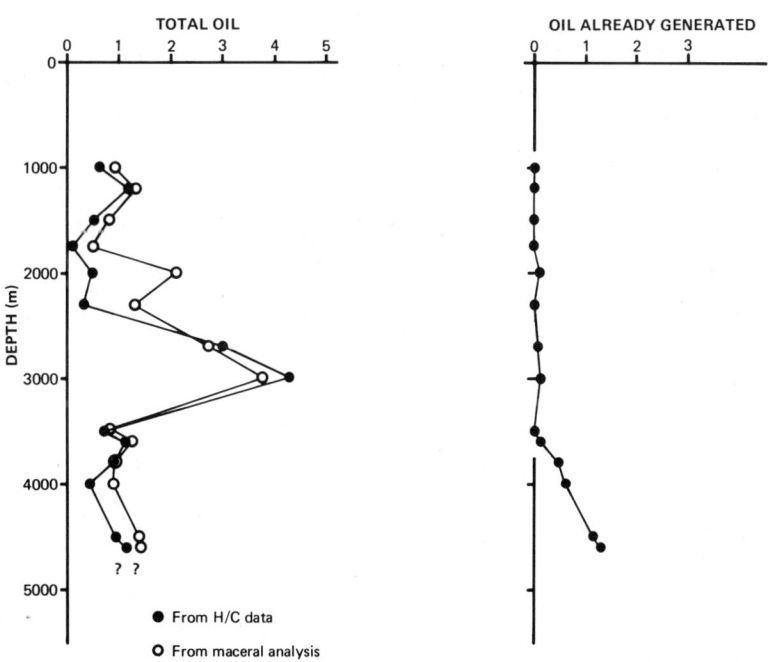

Figure 9.2. "Total Oil" and "Oil Already Generated" profiles for Mauve Well.

"Oil Already Generated" values indicate that only the section lying below 4500 meters is likely to have generated anything approaching a commercially attractive amount of oil. The relative organic richness of the blackened samples below 4600 meters makes them interesting for further investigation. Finally, a more thermally mature version of the rocks lying between 2700 and 3000 meters in the Mauve Well could already have generated very large quantities of oil. Future exploratory activity could include an attempt to find such a section.

Problem 3

Perform a source-rock evaluation of the section penetrated in the Turquoise Well, data for which are given in Table 9.4.

Table 9.4. Source-rock data for the Turquoise Well

Depth (ft)	Type of Sample	TOC[a]	Bit/TOC[b]	Atomic H/C	R_O[c]	TAI[d]	% Alginite + Exinite
3000	Cuttings	1.0	0.06	0.90	0.49	2–2.5	40
3500		0.8	0.06	0.85	0.52	2.5	30
4000		0.7	0.05	0.86	0.59	2.5	35
4500		0.9	0.08	1.02	0.65	2.5–3	40
5000		1.1	0.91	0.91	0.67	2.5–3	50
5500		2.3	0.66	1.25	0.88	2.5–3	80
6000		2.6	0.22	1.21	0.91	2.5–3	75
6500		4.1	0.51	1.17	1.00	2.5–3	75
7000		0.5	0.08	0.65	1.07	3.0	25
7500		0.3	0.08	0.71	1.27	3–3.5	40
8000		1.8	0.27	0.99	1.21	2.5–3	70
8500		1.7	0.18	1.03	1.26?[e]	2.5–3.5	80
9000		0.2	0.01	0.60	1.41?[e]	3.5	20
9500		0.4	0.03	0.51	1.33?[e]	3–3.5	15
10,000	↓	0.3	0.02	0.48	1.51	3.5	10

[a] TOC = Total Organic Carbon
[b] Bit/TOC = Bitumen/Total organic carbon
[c] R_O = Vitrinite reflectance
[d] TAI = Thermal Alteration Index
[e] ? indicates a poor histogram

Solution. Here there are abundant data, because six different analyses were performed. Quantity is given by TOC, quality is measured by H/C and maceral analysis, and maturity is determined by R_O, TAI, and Bitumen/TOC.

Table 9.5 shows quality factors calculated from each of the two quality parameters. The only significant discrepancies appear in the most alginite-rich samples; these discrepancies probably are not serious, because the quality factors calculated from both sets of data are very high. The discrepancies might arise from failure to distinguish between alginite and exinite in calculating the quality factor by maceral analysis. The best solution is probably to average the two quality factors for each sample, as is done in Table 9.5.

Of the three maturity parameters, Bitumen/TOC is by far the least precise. There is a large increase in bitumen between 5000 and 6500 feet, but the very high values suggest that much of the bitumen is not indigenous. Vitrinite reflectance values and TAI measurements both suggest that the oil-generative zone lies between 4500 and 7500 feet. Given the general agreement between TAI and R_O, and the uncertainty in several of the TAI measurements (presented as ranges instead of specific values), it would probably be wisest to use R_O for calculating maturity. Note that for three samples (8500–9500 feet) the vitrinite reflectance values are listed as "uncertain" because the histograms (not shown) are poor. In this case, however, the questionable values seem reasonable on the basis of comparison with the more reliable values at 8000 and 10,000 feet.

Both the TAI and R_O values show a slight reversal between 7500 and 8000 feet. This phenomenon could be real if a fault cut the section there, pushing a sequence of low maturity below another of higher maturity. Such reversals are not uncommon, but their putative occurrence should always be checked against geologic and micropaleontologic data. Such reversals, especially a small one like that in the Turquoise Well, can also be the result of experimental error and uncertainty. In the present case, however, the fact that both R_O and TAI concur is good evidence that the reversal is real.

Plots of "Total Oil" and "Oil Already Generated" are shown in Figure 9.3. The "Total Oil" plot calls attention to the two excellent source rocks lying in the 5000–7000 and 7500–8500 foot intervals. Actually, more samples are needed to define accurately the limits of these rich formations. Well-log data might be helpful in this regard, but it would also be advisable to obtain and analyze several samples in each of the following intervals: 5000–5500 feet, 6500–7000 feet, 7500–8000 feet, and 8500–9000 feet. Correct appraisal of the thickness of the rich section is essential for prospect evaluation.

Table 9.5. Quality Factors for Samples from the Turquoise Well

Depth (ft)	Quality factor (H/C)	Quality factor (maceral anal.)	Quality factor (average)
3000	0.60	0.8	0.7
3500	0.50	0.6	0.55
4000	0.56	0.7	0.65
4500	1.00	0.8	0.9
5000	0.73	1.0	0.85
5500	2.5	1.6	2
6000	2.5	1.5	2
6500	2.5	1.5	2
7000	0.50	0.5	0.5
7500	0.86	0.8	0.8
8000	2.5	1.4	2
8500	2.5	1.6	2
9000	0.60	0.4	0.5
9500	0.24	0.3	0.3
10,000	0.24	0.2	0.2

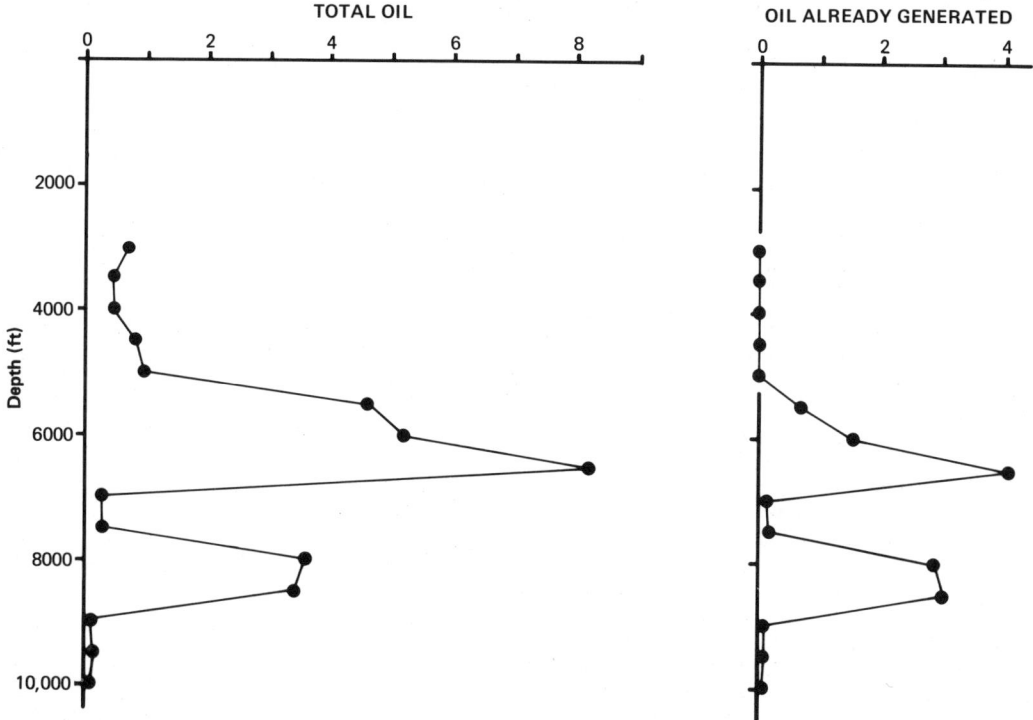

Figure 9.3 "Total Oil" and "Oil Already Generated" profiles for the Turquoise Well.

"Oil Already Generated" confirms that large amounts of bitumen have indeed been generated in these two intervals. The upper interval has not quite reached full maturity, but because of its high "Total Oil" values it has nevertheless been an effective oil generator. Future work in this basin should therefore focus on determining the areal extent and thermal maturity of the rich facies, the timing of oil generation, and the direction in which migration is likely to have occurred. Such a prospect is an exciting one.

But how do the bitumen contents fit into this source rock analysis picture? It was noted earlier that the very high bitumen contents in the 5000–6500 foot interval suggested nonindigenous bitumen. This could be the result of either contamination or migration. Contamination seems unlikely, because the high bitumen contents coincide with the rich part of the section, and it is unlikely that contamination would be limited to the good source rocks. It seems much more plausible that the bitumen is actually present in these rocks, although it could not have originated at precisely this point. Perhaps this source interval is sufficiently fractured to be functioning also as a reservoir for bitumen which has migrated up from downdip. This hypothesis should be carefully tested, for it may have significant implications for exploration. Careful analysis of the composition of the bitumen would be the next order of business.

Problem 4

Perform a source-rock analysis on the samples from the Gold Well, data for which are given in Table 9.6.

Table 9.6. Source-Rock Data for Samples from the Gold Well

Depth (ft)	Type of sample	BFOC[a]	Atomic H/C	TAI[b]	Bit/BFOC[c]
500	Cuttings	3.1	1.27	2.0	0.15
600	↓	3.7	1.15	2.0	0.12
800		2.6	1.21	2–2.5	0.13
1000		2.9	1.08	2–2.5	0.16
1200		1.8	0.95	2.0	0.07
1500		3.1	1.35	2.0	0.18
2000	↓	2.7	1.41	2–2.5	0.15
2500	Core	3.0	1.27	2.5	0.07
3000	Cuttings	4.6	1.22	2–2.5	0.13
3500		1.7	1.27	2–2.5	0.15
4000		1.9	1.33	2.3	0.12
4500		1.6	1.15	?	0.13
5000		1.9	1.19	?	0.19
5500		2.3	0.87	2.0–3.5	0.17
6000	↓	2.7	1.22	2.5	0.16

[a]BFOC = Bitumen-free organic carbon
[b]TAI = Thermal Alteration Index
[c]Bit/BFOC = Bitumen/Bitumen-free organic carbon

Solution. This analysis is quite straightforward, except for a few minor problems. Quantity is measured by BFOC, quality by H/C ratio, and maturity by TAI. Of these, only TAI presents any difficulties, because most samples in the Gold Well are assigned a range of TAI values rather than a single, exact one. This indicates an uncertainty on the part of the microscopist, and normally would suggest a need for vitrinite reflectance analyses for better accuracy. In the present case, however, there seems to be little doubt that the entire section is immature, so reflectance work would probably be a waste of time and money.

Plots of "Total Oil" and "Oil Already Generated" are shown in Figure 9.4. Although the entire section has thus far not generated any oil, in a deeper, hotter setting much of it could be an excellent source rock.

A problem remains, however. Why are the Bitumen/BFOC values so high in most of the samples? Immature sections usually have Bitumen/BFOC values less than 0.1.

The answer here probably lies in sample quality. Note that all the samples except one are cuttings, the least reliable type of sample. The single core sample has a lower bitumen content than the cuttings. These data suggest (but do not prove) that some systematic contamination of the cuttings samples has occurred, and that bitumen analyses of these samples are likely to be worthless. This example demonstrates how a careful analysis of preliminary data can save time and money by showing which future analyses are likely to be helpful, and which are not.

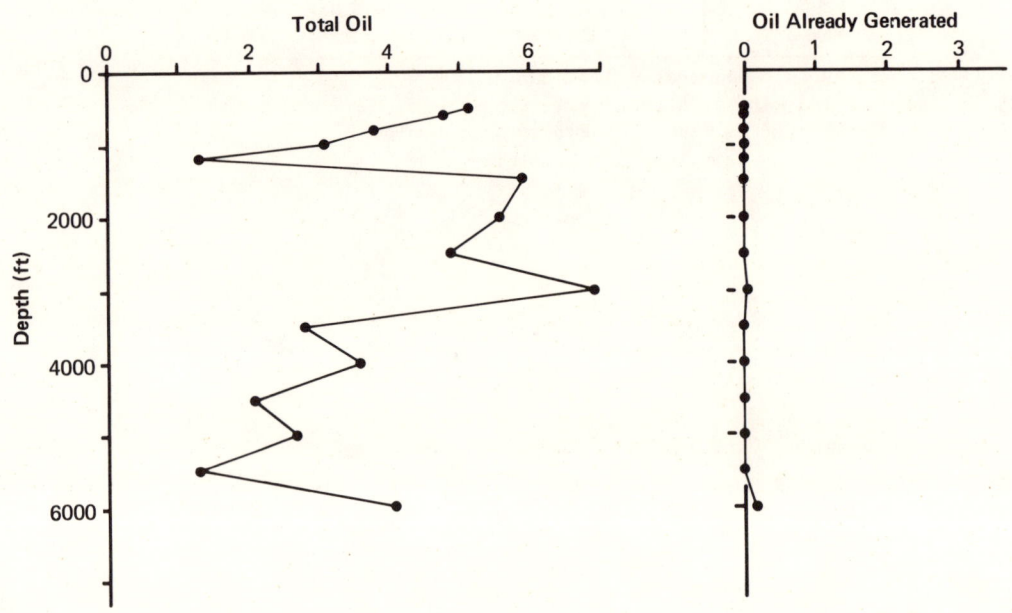

Figure 9.4. "Total Oil" and "Oil Already Generated" profiles for the Gold Well.

2. Oil–Oil Correlations

Problem 5

Three light oils are obtained from slightly different depths in three wells located about a mile apart in the Kangaroo Field. Determine whether all three oils are from a common source. Data for the three oils are given in Table 9.7.

Table 9.7. Data for Kangaroo Field Oils

	Purple Well	Scarlet Well	Ochre Well
Depth (ft)	8750–8780	10,312–10,408	9592–9666
Age of reservoir	Miocene	Eocene	Eocene
API gravity (°)	48.3	46.7	36.0
% Sulfur	0.08	0.07	0.09
CPI_{23-31}	1.05	1.03	1.02
Porphyrins			
Ni	0.00	0.00	0.00
V	0.01	0.01	0.02
$\delta^{13}C$ (whole crude) vs PDB	−24.9	−25.2	−25.1
$\delta^{13}C$ (sat. HC) vs PDB	−26.3	−26.1	−26.0
Saturated HC (% of whole crude)	48.9	50.7	38.2
Aromatic HC (% of whole crude)	10.2	19.6	19.1
Pristane/Phytane	1.27	1.33	1.15

Solution. A quick look at the data reveals no significant differences between the oils from the Purple Well and the Scarlet Well. The close similarities of many unrelated parameters (porphyrins, carbon isotopes, hydrocarbons, sulfur content) strongly suggest that these two oils are virtually identical. The simplest explanation is that they are from a common source. As was noted in Chapter 7, however, these data cannot prove conclusively that the two oils are genetically related. Absolute proof is beyond our capability, but the present example offers proof "beyond a reasonable doubt."

The oil from the Ochre Well, however, is slightly different from the other two. All of the measured properties except API gravity and saturated hydrocarbon content are virtually identical with those of the Purple and Scarlet oils. The fact that such diverse parameters as carbon isotopes, porphyrins, and isoprenoid hydrocarbons are in agreement indicates a strong possibility that there is a genetic relationship among all three oils. If all are from a common source, migrational differences might account for the compositional differences.

Its lower gravity suggests that the Ochre oil has had an easier migration. Such factors as length of migration, especially primary migration, and faulting could have influenced the degree to which composition of the oil changed during migration. It might be proposed as a working hypothesis that the common source for all three oils is physically closest to the Ochre oil.

As a final confirmation of the proposed genetic relationship among the three oils, gas chromatograms of the oils (either of the whole oils or of the saturated hydrocarbon fractions) should be obtained and compared peak by peak. (Gas chromatography is very useful for fingerprinting crude oils and bitumens in this way.)

Problem 6

Two oil reservoirs are penetrated by the Gray Well. One is located at about 3000 feet, and the other at 9000 feet. Determine whether the two oils, which have very different appearances, could have a common source. Data for the oils are presented in Table 9.8, and chromatograms of the oils are shown in Figure 9.5.

Table 9.8. Data for Gray Well Oils

		3000' oil	9000' oil
$\delta^{13}C$ (whole oil, vs PDB)		−23.3	−22.7
API gravity		25°	12°
% Sulfur		0.18	0.04
Porphyrins	Ni	<0.02	0.12
	V	<0.02	0.26

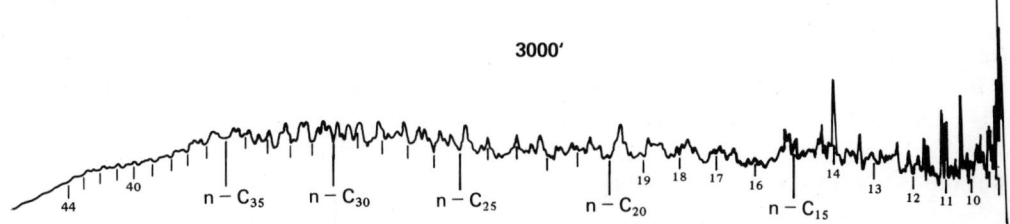

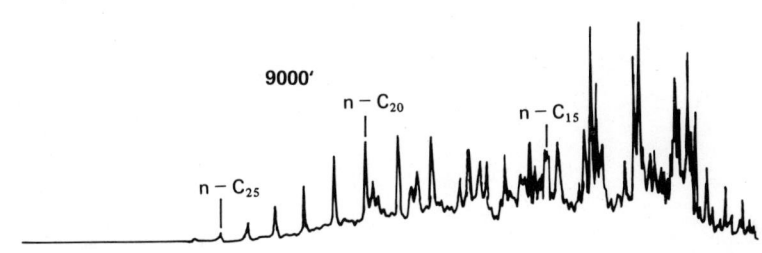

Figure 9.5. Gas chromatograms of the Gray Well oils.

Solution. It is clear that the two oils are now vastly different from each other. One important clue to the difference comes from the gas chromatogram for the 3000′ oil. The absence of *n*-paraffins strongly suggests that biodegradation has occurred. If the two oils did have a common origin, microbial transformations could have been responsible for the increased sulfur content of the degraded crude. On the other hand, biodegration should decrease the specific gravity, increase porphyrin content, and make the $\delta^{13}C$ value less negative. All of these expected trends are contrary to what actually is observed.

Migrational differences could explain the higher gravity, lower porphyrin content, and more negative carbon-isotope value for the 3000′ oil. Because migration usually occurs upward, if the two oils had a common origin, the 3000′ oil would have migrated farther than the 9000′ oil.

At this point, therefore, one might be tempted to say that although there is no specific evidence which indicates that the two oils are genetically related, there are no data which conclusively contradict the hypothesis. The chromatograms, however, contain more evidence. If one ignores the differences in *n*-alkanes caused by biodegradation, and looks instead at the other components, it is seen that there is little correspondence between the two oils. For example, the two peaks of moderate size that appear between $n\text{-}C_{18}$ and $n\text{-}C_{19}$ in the 9000′ oil are not present in the 3000′ oil. In the 3000′ oil there is a single dominant compound to the right of $n\text{-}C_{15}$, while in the 9000′ oil there are several peaks. Many other differences are apparent, and it is unlikely that a combination of biodegradation and migration could explain them. One would be justified therefore in concluding on the basis of the greatly different chromatograms that the two oils do not have a common source.

Problem 7

The South Hootchiekootchie Basin is a prolific oil-producing region, with many offshore wells. One day a large oil slick is noted exactly equidistant from 12 producing wells. None of the well operators will confess to having had a leak in his system, but the Department of Interior demands satisfaction. Your job is to identify the culprit by using organic geochemistry.

You first request fresh samples of oil from each of the producing wells, and from the least-weathered part of the oil slick. You then request that the following analyses be carried out on each of the samples: gas chromatography of saturated hydrocarbons, carbon isotope ratios on the topped crudes, porphyrin contents, sulfur contents, and API gravities. Results of these analyses are given in Table 9.9. Are these data sufficient to definitively correlate the spilled oil with a single well?

Table 9.9. Analytical Data on Spilled Oil and Twelve Possible Sources for the Spill

Well	Production Depth (ft)	Producing Formation	API Gravity (°)	% Sulfur	$\delta^{13}C$, topped oil (vs PDB)
1	8100–8150	C	30.5	0.8	–26.7
2	6763–6849	A	27.5	0.3	–29.7
3	8460–8610	C	34.5	0.1	–27.9
4	7998–8016	C	19.7	0.7	–26.7
5	9001–9202	C	31.8	1.3	–28.0
6	6887–6990	C	29.7	1.2	–28.1
7	7421–7503	C	25.7	1.7	–26.4
8	9023–9112	B	24.3	2.2	–26.2
9	6810–7020	A	28.1	0.1	–30.1
10	9100–9140	D	36.1	0.2	–28.4
11	8321–8520	D	24.3	1.7	–27.2
12	8169–8230	C	21.2	0.6	–26.9
Spill Oil	–	–	12.7	3.7	–25.7

Well	Porphyrins Ni	Porphyrins V	Pristane / n-C_{17}	Phytane / n-C_{18}	Pris / Phyt	CPI_{23-31}	Maximum n-paraffin (C#)
1	1.02	1.55	0.81	0.68	1.2	1.01	17
2	0.00	0.03	0.03	0.43	0.1	1.13	18
3	0.16	0.12	0.23	0.92	0.2	1.02	17
4	1.21	1.02	0.21	0.67	0.3	1.04	16
5	0.92	1.45	1.02	0.88	1.2	0.99	17
6	1.02	0.34	0.92	0.91	1.0	1.07	17
7	0.99	0.29	0.19	0.66	0.3	1.04	17
8	0.15	0.05	0.31	1.72	0.2	0.87	18
9	0.02	0.00	0.11	0.51	0.2	1.17	17
10	0.15	0.13	0.49	0.96	0.5	1.06	18
11	0.88	0.21	0.97	0.79	1.2	1.03	16
12	1.15	0.96	0.16	0.81	0.2	0.98	17
Spill Oil	0.72	0.27	3.65	16.1	0.2	–	–

Solution. The spilled oil has probably undergone a significant amount of evaporation and biodegradation as a result of exposure to sea water and air. API gravity should therefore not be very meaningful, nor should *n*-paraffin distributions, because biodegradation removes *n*-alkanes preferentially. The pristane/phytane ratio, however, may not have been affected, because branched hydrocarbons are not attacked as readily as are straight-chain ones. Sulfur content would probably have risen slightly during biodegradation. The carbon isotopic composition would have become slightly biased toward the heavier (^{13}C) isotope as a result of the loss of isotopically light alkanes and light aromatics. Porphyrin content would have increased as a consequence of the preferential loss of hydrocarbons by evaporation and microbial degradation. The Ni/V ratio, however, should not have been affected.

By putting all these facts together, one can evaluate the data in Table 9.9; results of these correlations are shown in Table 9.10. All data from the well samples that are definitely incompatible with those of the spill oil are set in boldface, and all dubiously correlated data are set in italic.

The pristane/phytane ratio of the spilled oil is about 0.2. All pristane/phytane ratios greater than 1.0 are impossibly high, and oils that have these ratios can be eliminated. Oil 10, with a ratio of 0.5, also is probably not the source. Note that the very high pristane/*n*-C_{17} and phytane/*n*-C_{18} ratios are almost certainly the result of biodegradation after the spill, and are not useful for correlation.

The Ni/V ratio in the spilled oil is about 3, and so a number of oils whose Ni/V ratios are not near 3 can be eliminated. Oils 2 and 9 are probably not the source because their porphyrin contents are too low. Of the remaining four oils, numbers 6, 7, and 11 all have more porphyrins than does the spill oil. If one of them were the source, porphyrin content would have decreased after the spill even faster than did evaporation and biodegradation. This seems unlikely, but not totally impossible, because extensive water washing might remove the relatively soluble porphyrins to a significant degree.

Sulfur content of the spilled oil is higher than in any of the possible source oils. Although sulfur will be concentrated as a consequence of the loss of sulfur-free hydrocarbons, there are limits to how much the sulfur content could be increased by concentration effects alone. Certainly the oils which are very low in sulfur (those that contain less than 0.5%) could not have been the source. Oils 1, 4, and 12, with 0.8, 0.7, and 0.6% sulfur, respectively, could possibly qualify.

Carbon isotopes are also useful. It is known that the various fractions of a crude oil usually do not differ by more than about 2 ‰ in δ^{13}C value, so if the spilled oil had lost all of its isotopically light saturated hydrocarbons, its overall δ^{13}C value could not have changed by more than 2 ‰. Because the spill oil had δ^{13}C = –25.7 ‰ versus PDB, the **source oil probably had δ^{13}C between –25.7 and –27.7 ‰**.

It is evident from Table 9.10 that only two oils, numbers 7 and 8, are possible sources for the spill. Of these, oil 8 gives a closer fit for all of the parameters, but there is not a clear choice between the two.

At this point it would be necessary to carry out some further "fingerprinting" analyses to identify the spilled oil as one of the two remaining suspects. It would be necessary to look at some parameter which could not have been affected by water washing, evaporation, or biodegradation. It would be appropriate to look at the steranes and triterpanes present in all three oils as a means of fingerprinting them. The next analytical step, therefore, would be to take the branched-cyclic fraction of the saturated hydrocarbons and subject it to gc–ms analysis, as Seifert and Moldowan (1979) have done.

Table 9.10. Analytical Data on Spilled Oil and Twelve Possible Sources for the Spill

Well	Production Depth (ft)	Producing Formation	API Gravity (°)	% Sulfur	$\delta^{13}C$, topped oil (vs PDB)
1	8100–8150	C	30.5	0.8	−26.7
2	6763–6849	A	27.5	0.3	−29.7
3	8460–8610	C	34.5	0.1	−27.9
4	7998–8016	C	19.7	0.7	−26.7
5	9001–9202	C	31.8	1.3	−28.0
6	6887–6990	C	29.7	1.2	−28.1
7	7421–7503	C	25.7	1.7	−26.4
8	9023–9112	B	24.3	2.2	−26.2
9	6810–7020	A	28.1	0.1	−30.1
10	9100–9140	D	36.1	0.2	−28.4
11	9321–8520	D	24.3	1.7	−27.2
12	8169–8230	C	21.2	0.6	−26.9
Spill Oil	—	—	12.7	3.7	−25.7

Well	Porphyrins Ni	Porphyrins V	Pristane n-C_{17}	Phytane n-C_{18}	Pris/Phyt	CPI_{23-31}	Maximum n-paraffin (C#)
1	1.02	1.55	0.81	0.68	1.2	1.01	17
2	0.00	0.03	0.03	0.43	0.1	1.13	18
3	0.16	0.12	0.23	0.92	0.2	1.02	17
4	1.21	1.02	0.21	0.67	0.3	1.04	16
5	0.92	1.45	1.02	0.88	1.2	0.99	17
6	1.02	0.34	0.92	0.91	1.0	1.07	17
7	0.99	0.29	0.19	0.66	0.3	1.04	17
8	0.15	0.05	0.31	1.72	0.2	0.87	18
9	0.02	0.00	0.11	0.51	0.2	1.17	17
10	0.15	0.13	0.49	0.96	0.5	1.06	18
11	0.88	0.21	0.97	0.79	1.2	1.03	16
12	1.15	0.96	0.16	0.81	0.2	0.98	17
Spill Oil	0.72	0.27	3.65	16.1	0.2	—	—

3. Oil–Source Rock Correlations

Problem 8

Apparent oil staining was detected in a sandstone core taken at 7927 feet in the Lavender Well. What is the nature and origin of this organic material? Source-rock data for the Lavender Well are given in Table 9.11.

Table 9.11. Source-Rock Data for the Lavender Well

Depth (ft)	TOC	% Alginite + Exinite	TAI
1000	1.2	60	1.5–2
1500	1.5	70	2.0
2000	1.8	60	2.0
2500	0.9	50	2.2
3000	1.3	70	2.3
3500	2.6	80	2.2
4000	2.1	80	2.5
4500	1.5	75	2.5
5000	1.2	50	2.5
5500	1.3	35	2.6
6000	1.9	50	2.5
6500	1.0	25	2.6
7000	0.5	10	2.7
7500	0.8	10	2.7
8000	1.3	60	2.9
8500	1.4	80	2.8
9000	1.1	70	3.0
9500	3.7	90	2.7
10,000	3.2	80	3.0
10,500	1.3	60	3–3.5
11,000	0.1	100	3.5
11,500	0.2	95	3.5
12,000	0.1	95	3.5
12,500	0.4	90	3.5

Solution. The oil-source history of the section can be evaluated from the quantity, quality, and maturity data in Table 9.11. The TAI data in the table show some scatter, particularly in the oil-generative zone (TAI 2.6 to 3.2), so it would be wise to plot the TAI data against depth and then obtain the best fit of the maturity curve to all the data. This is done in Figure 9.6.

The maturity line of Figure 9.6 can be used to calculate "Total Oil" and "Oil Already Generated" for the complete section; the results of such a calculation are shown in Figure 9.7. It is readily apparent that the most likely local source for the oil in the 7927' core is the section that lies between 9000 and 10,500 feet. This interval lies structurally below the stain, so upward migration could have occurred. Also, the rocks in this interval have already generated large amounts of bitumen.

The data in Table 9.12 indicate that the staining in the 7927' core is not a refined petroleum, because the porphyrin content is high (porphyrins are destroyed by refining), and because there is a large amount of heavy *n*-paraffins present.

The staining does not appear to come from the 9000' sample, because the correlations of porphyrin contents, carbon-isotope ratios, and *n*-paraffin distributions are all poor. Similarly, the 10,500' sample cannot be the source, because the carbon-isotope ratio of the 10,500' kerogen is more negative than that of the oil stain. The 10,500' kerogen should have produced a bitumen with a $\delta^{13}C$ value of -29 to -30 ‰ vs. PDB, and migration from the 10,500' level to 7927' would make the $\delta^{13}C$ value even more negative.

Most of the bitumen present in the 10,500' sample cannot be indigenous to that sample because the isotope values of kerogen and bitumen are incompatible. This bitumen must have migrated into the 10,500' horizon. Its most probable origin is the sediments lying stratigraphically above it in the 9500–10,000' interval. There is a generally satisfactory correlation between these bitumens and the 10,500' bitumen. Prophyrin contents are slightly anomalous, but this may be the result of a minor contribution from indigenous bitumen which was very rich in porphyrins.

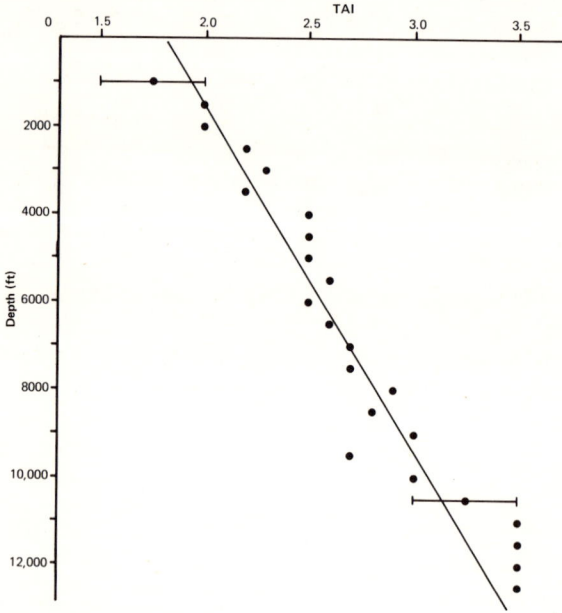

Figure 9.6. Thermal maturity curve for the Lavender Well.

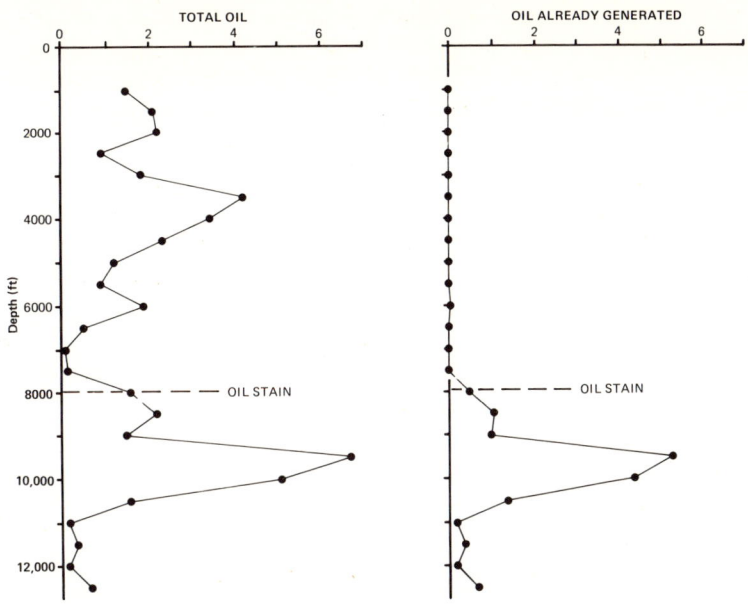

Figure 9.7. "Total Oil" and "Oil Already Generated" for the Lavender Well.

Table 9.12. Analytical Data for Oil Stain and Bitumens from Postulated Source-Rock Intervals, Lavender Well.

| Depth (ft) | Porphyrins | | $\delta^{13}C$ (‰ PDB) | | $\dfrac{Pristane^{a}}{Phytane}$ | CPI^{a}_{23-31} | maximum* |
	Ni	V	Kerogen	Bitumen			n-paraffin
7927	1.21	0.55	—	−26.0	0.57	1.09	C_{25}
9000	0.88	0.15	−30.2	−30.7	0.92	1.02	C_{17}
9500	1.52	0.59	−26.1	−27.2	0.48	1.18	C_{27}
10,000	1.37	0.65	−25.3	−25.7	0.66	1.29	C_{27}
10,500	1.91	1.02	−28.3	−25.5	0.51	1.28	C_{27}

[a]obtained from gas chromatograms (not shown)

The fact that bitumen from the 9500–10,000′ interval has migrated into the 10,500′ interval is a strong indication that upward migration into the 7927′ core has also occurred. Indeed, there is an excellent correlation among all the measured parameters from the oil stain and the bitumens. Before making a final pronouncement that these samples are genetically related, however, it would be good practice to perform some further, more sophisticated tests, such as comparing the gas chromatograms in detail (fingerprinting), measuring the carbon-isotope ratios of the saturated hydrocarbon, aromatic hydrocarbon, resin, and asphaltene fractions, and looking at certain steranes and triterpanes as Seifert and Moldowan (1979) have done.

Problem 9

The Green Well is drilled 90 miles northwest of the Mauve Well (Problem 2), and an oil reservoir is penetrated in a Pliocene sand at a depth of about 1200 meters. Because both the Mauve and Green Wells are wildcats, there are no rock samples or data available for the region except those already presented in Tables 9.2 and 9.3, and Figures 9.1 and 9.2. Could there be a genetic relationship between the Green oil and the strata analyzed in the Mauve Well?

Solution. As shown in Figure 9.2, there are two intervals in the Mauve Well that should be considered as possible oil source rocks: 2500–3500 meters and 4500–4600 meters. One should therefore attempt to compare bitumens from these intervals with the Green oil. Analyses which are likely to be of value are carbon isotope ratios, porphyrin contents, sterane and triterpane contents, and *n*-paraffin and isoprenoid distributions. These data are presented in Table 9.13.

Table 9.13. Analytical Data for Green Oil and Mauve Bitumens

Well	Depth (m)	$\delta^{13}C$, ‰ (vs. PDB)	Porphyrins Ni	Porphyrins V	*n*-paraffin distribution	Pristane/Phytane	C_{29} steranes/Cholestane
Green	1200	−29.3 oil	0.11	0.15	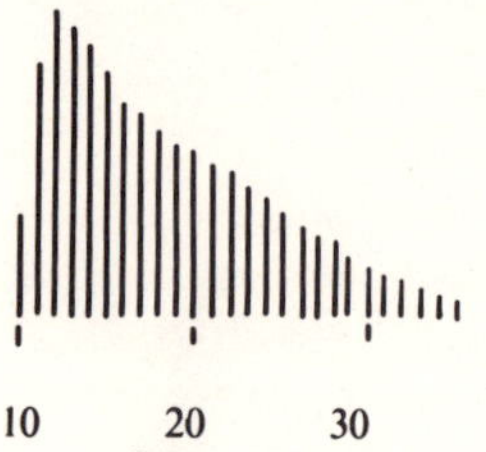	1.15	0.34
Mauve	2700	−28.5 bit. / −27.4 ker.	0.42	0.44	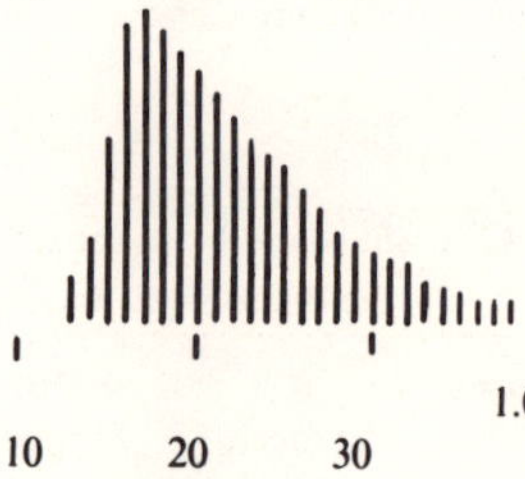	1.00	0.41
Mauve	3000	−26.3 bit. / −26.1 ker.	0.21	0.60	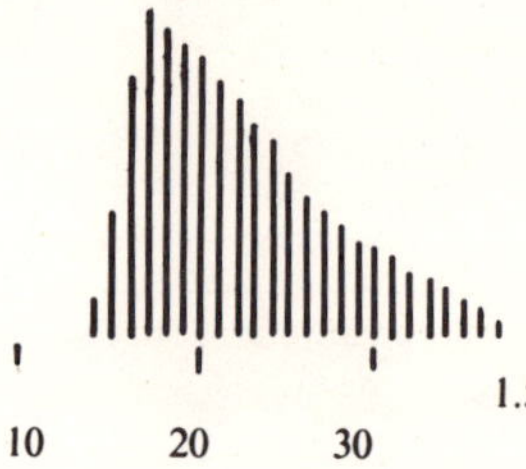	1.51	0.19
Mauve	4500	−24.7 bit. / −23.8 ker.	0.72	0.35	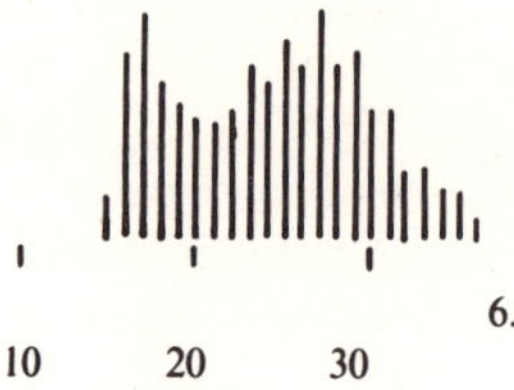	6.41	1.12
Mauve	4600	−25.5 bit. / −24.4 ker.	0.41	0.52	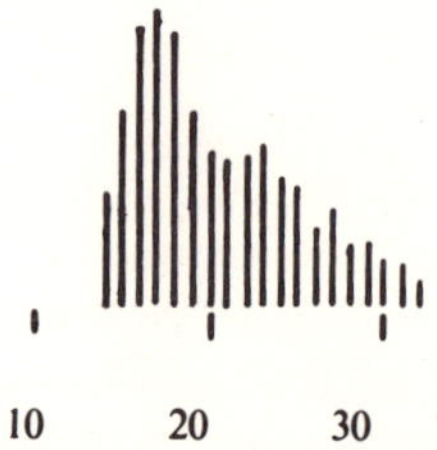	5.28	0.87

On the basis of these data, it can be concluded that the rocks in the 4500–4600-meter interval cannot have been the source of the Green oil. ^{13}C to ^{12}C ratios for the kerogens from these rocks are 5 to 6 °/oo greater than these for the Green oil. This difference is too large to have been caused by thermal effects or migration. Furthermore, pristane/phytane ratios and sterane contents are totally different, and the *n*-paraffin distributions also differ substantially. The deep samples from the Mauve Well have relatively large amounts of heavy homologs and high CPI values, pointing to a significant contribution from terrestrial waxes.

The 2700–3000 meter samples, however, show a much closer correlation. Carbon isotope values are more or less compatible, differing by 1.9 and 3.1 °/oo for kerogens and oil, respectively. Porphyrin contents in the 2700-meter sample correlate reasonably well when migrational loss is taken into account, but those in the 3000-meter sample do not. Pristane/phytane ratios are acceptably close for all three. Sterane contents and *n*-paraffin distributions are also similar.

We conclude therefore that the rocks found in the 2700-meter interval in the Mauve Well correlate well with the Green oil. The 3000-meter rocks do not correlate as well, but the differences are not large enough to completely rule out a genetic relationship.

But what do these results mean in terms of exploration? It is obvious that the rocks analyzed in the Mauve Well could not have been the actual source rocks for the Green oil, for two reasons. Firstly, as shown in Figure 9.2, they are thermally immature and have not yet generated substantial amounts of bitumen. Secondly, the distance between the two wells (90 miles) is rather large.

We cautiously extend the hypothesis that the Green oil may have had its source in a lateral equivalent of the 2700-meter (and/or 3000-meter) rocks in the Mauve Well. This lateral equivalent would have to be similar in facies and organic geochemistry to the Mauve samples, but would have to have been subjected to greater thermal maturation, most likely as the result of deeper burial. The lateral equivalent would also ideally be much closer to the Green oil reservoir.

This hypothesis assumes that the Green oil is attributed to a source rock whose existence has not yet been verified. Future work should attempt to determine the lateral extent of the interesting strata present in the Mauve Well, and to ascertain whether the entire region is of the same oil-source capacity as the Mauve samples. A clear determination of the volume of source rock will in turn enable the explorationist to evaluate the oil-source potential of the region.

4. Lopatin Reconstructions

Problem 10

The Black Well was drilled off the Louisiana Gulf Coast. It penetrated 1000 feet of Pleistocene sediments, 3500 feet of Pliocene rock, and 11,000 feet of Upper Miocene before being abandoned at 16,150 feet in the Middle Miocene. The bottom hole temperature was found to be 270 °F. Make a Lopatin reconstruction for the well and calculate the TTI at total depth.

132   *Organic Geochemistry for Exploration Geologists*

Solution. With only a bottom hole temperature and no information with which to reconstruct a fancy temperature history of this section, it would be best to assume that the present-day geothermal gradient is constant through the section, and that it has remained constant since the Middle Miocene. If a present-day surface temperature of 68 °F is assumed, the gradient was 202 °F/16,150 feet, or 1.24 °F/100 feet. After converting Fahrenheit to Celsius and inverting, one finds a thickness of 1442 feet for every 10 °C interval. Reconstruction can therefore be begun by placing the isotemperature lines on the time–depth plot, as shown in Figure 9.8.

Next the sediment horizon lines are added. The data permit a complete definition of three horizons: the Pleistocene–Pliocene boundary, the Pliocene–Upper Miocene boundary, and the boundary between Upper and Middle Miocene. By taking the dates for these three boundaries as 1.8, 5, and 12 million years ago, respectively, the three horizon lines shown in Figure 9.9 are obtained. A fourth horizon line corresponding to the sediment at total depth can also be obtained, but its exact age is unknown and the line cannot be completed all the way to the surface. Fortunately, however, the temperature is so low near the surface that this uncertainty is unimportant.

Calculation of the TTI at the bottom of the hole is now easy. With the use of the temperature factors given in Table 8.1, and estimates of the time spent in each temperature interval, the TTI is calculated to be 17.69 at 16,150 feet.

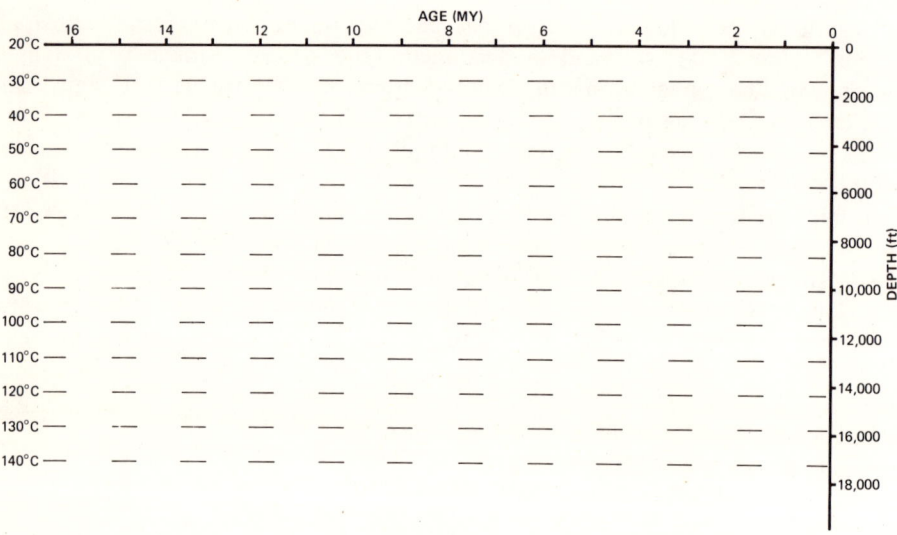

Figure 9.8. Subsurface temperature grid for the Black Well.

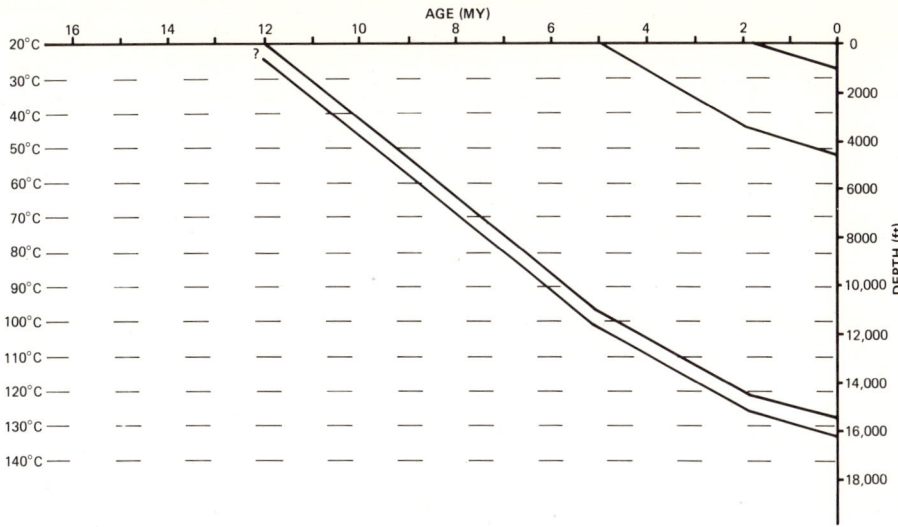

Figure 9.9. Lopatin reconstruction for the Black Well.

Problem 11

The Ultraviolet Well is spudded in Paleocene sediments. At a depth of 1500 feet, micropaleontology indicates that the rocks are of Maastrichtian age. The following Upper Cretaceous boundaries are noted:

Maastrichtian–Campanian	1807 feet
Campanian–Santonian	2002 feet
Santonian–Coniacian	2360 feet
Coniacian–Turonian	2546 feet
Turonian–Cenomanian	3017 feet

The Cenomanian is 480 feet thick and overlies 1000 feet of Kimmeridgian shale. Total depth is reached at 6120 feet in Middle Jurassic rocks.

Evidence from related sections indicates that the Paleocene was originally about 3000 feet thick, and that no other Cenozoic sediments were ever deposited. Total original thickness of the Kimmeridgian is thought to be 1500 feet. It is also believed that 500 feet of Lower Cretaceous sediments were deposited before uplift and erosion began.

Assuming a surface temperature of 10 °C and an geothermal gradient of 2 °F/100 feet, draw a Lopatin reconstruction for the section. Show also the thermal history of the Kimmeridgian.

Solution. Draw the temperature grid as in the previous problem after converting Fahrenheit to Celsius; then consider the problem of reconstructing the horizon lines when unconformities are present.

The first unconformity is at the surface, where Paleocene sediments are exposed. Loss of 1500 feet of Paleocene sediment has occurred in the last 55 million years. In the absence of any evidence to the contrary, it can be assumed that erosion has occurred at a constant rate throughout this time interval. The temperature and depth of burial therefore have also decreased at a constant rate.

The second unconformity is also erosional. The section has lost 500 feet of Kimmeridgian and 500 feet of Lower Cretaceous. The exact time interval represented by the lost Lower Cretaceous sediments is not known, but this lack of knowledge is not serious, because all these events occurred when the bed of interest (the Kimmeridgian) was at very low temperatures.

Uplift in a basin is usually preceded by a slowing of the depositional rate. Assume that the Lower Cretaceous sediments were deposited at a slower rate than the Kimmeridgian. In this case a rate of 500 feet in 10 million years was selected. If it is also assumed that erosion occurred at a constant rate between 135 mybp and 100 mybp, the horizon lines can be completed as shown in Figure 9.10. The dashed line represents the top of the uneroded Kimmeridgian.

The TTI values can be calculated for the bottom and top of the uneroded Kimmeridgian. Note that the dashed line was followed in the early stages of deposition, because this is the top of the uneroded section. None of the Kimmeridgian turns out to be thermally mature; TTI values lie between 2.7 and 4.9, and therefore correspond to TAI and R_o values of about 2.5 and 0.5 to 0.55, respectively (Waples, 1980).

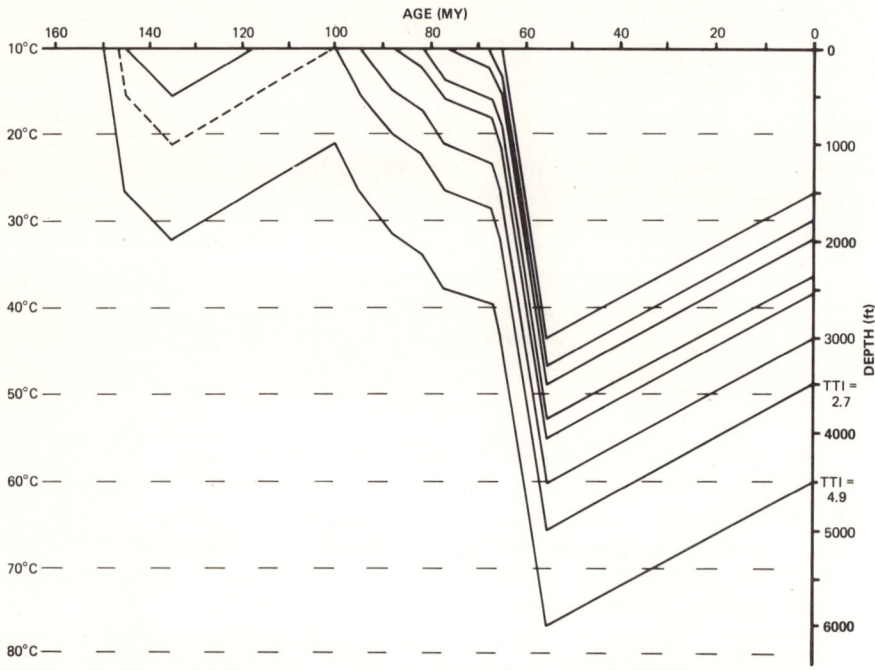

Figure 9.10. Lopatin reconstruction for the Ultraviolet Well.

Problem 12

Analyze the timing of oil generation in the Pink Well, drilled in the Midcontinent (U.S.) region. Data are given in Table 9.14. The geothermal gradient was found to be $1.0\,°F/100$ feet.

Table 9.14. Stratigraphy of Pink Well

Formation	Age (my)	Period	Depth (ft)
top-Permian	230	Perm.	0
top-Virgil	280	Penn.	7000
top-Missourian	288	Penn.	8000
top-Des Moines	296	Penn.	11,000
top-Atoka	304	Penn.	13,000
top-Morrow	309	Penn.	18,500
top-Mississippian	320	Miss.	21,000
top-Kinderhook	340	Miss.	23,000
top-Sylvan	425	Ord.	25,500
top-Arbuckle	470	Ord.	27,500

Solution. Some assumptions must be made about erosion since deposition of the Virgil. It might be reasonable to assume erosion totalling 2000 feet, at a constant rate through the past 280 million years. Let us also assume an average surface temperature of 15 °C (59 °F). The resultant Lopatin reconstruction is shown in Figure 9.11.

To draw iso-maturity lines on this reconstruction, one must find the points on each horizon line which correspond to TTI values of 15 and 160. These limits define the oil-generation window (Table 8.2). Figure 9.12 shows the iso-maturity lines superimposed on the reconstruction.

It can quickly be determined from Figure 9.12 that at the present time, the interval from 7600 feet to 14,000 feet is in the oil-generation zone. Temperatures in this interval range from 58 to 93 °C (136 to 199 °F). The relatively low temperatures in the present-day oil-generative window are a direct consequence of the long time that these sediments have spent in the subsurface.

In past epochs, the oil-generative window was deeper and hotter, because the sediments had not been cooked so long. The unusually high temperatures (115–140 °C) and great depths of burial (18,000–22,500 feet) required for oil generation from the Sylvan Formation were a result of the rapidity with which the sediments were deposited.

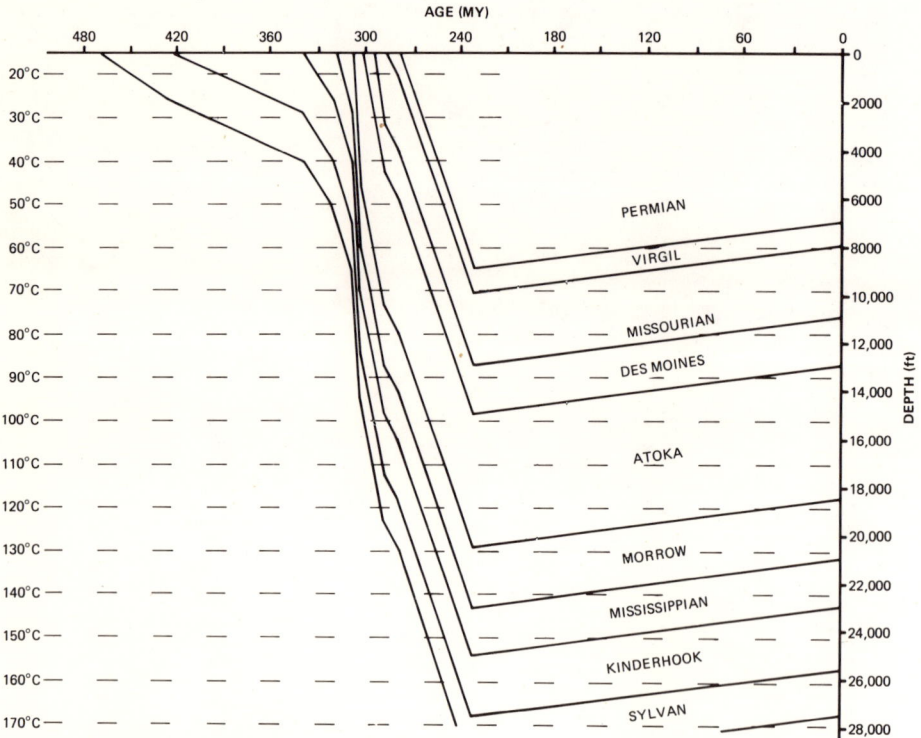

Figure 9.11. Lopatin reconstruction for the Pink Well.

Practice Problems

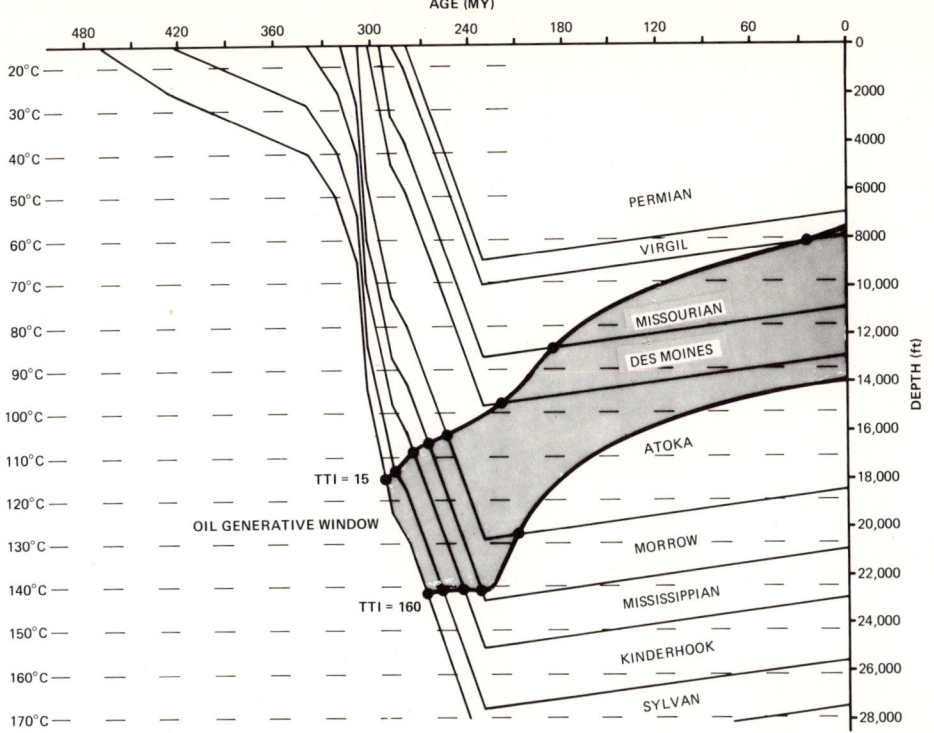

Figure 9.12. Iso-maturity lines superimposed on Lopatin reconstruction for the Pink Well.

REFERENCES

Albrecht, P. and Ourisson, G., 1969, Diagenesis of the saturated hydrocarbons in a thick sedimentary sequence (in French): Geochim. Cosmochim. Acta, 33:138-142.

Albrecht, P., Vandenbroucke, M., and Mandengue, M., 1976, Geochemical studies on the organic matter from the Douala Basin (Cameroon). I. Evolution of the extractable organic matter and the formation of petroleum: Geochim. Cosmochim. Acta, 40:791-799.

Alexander, R., Kagi, R. I., and Woodhouse, G. W., 1979, A new method for measuring the maturity of petroleum in source rocks: APEA Journal, 19:90-93.

Bailey, N. J. L., Krouse, H. H., Evans, C. R., and Rogers, M. A., 1973, Alteration of crude oil by waters and bacteria—evidence from geochemical and isotope studies: Bull. Am. Assoc. Petr. Geol., 57:1276-1290.

Baker, E. G., 1959, Origin and migration of oil: Science, 129:871-874.

―――, 1962, Distribution of hydrocarbons in petroleum: Bull. Am. Assoc. Petr. Geol., 46:76-84.

Baker, E. W., 1966, Mass spectrometric characterization of petroporphyrins: J. Am. Chem. Soc., 88:2311-2315.

Bazhenova, O. K., and Gorshkov, B. I., 1973, Characteristics of the distribution of organic material in Paleogene-Neogene formations of eastern Kamchatka (in Russian): Moscow University Vestnik, Ser. IV, Geology, No. 6, pp. 80-85.

Bernard, B. B., Brooks, J. M., and Sackett, W. M., 1976, Natural gas seepage in the Gulf of Mexico: Earth Planet. Sci. Lett., 31:48-54.

Bolin, B., 1970, The carbon cycle: Scientific American, 233(3):125-132.

Bonham, L. C., 1980, Migration of hydrocarbons in compacting basins: Bull. Am. Assoc. Petr. Geol., 64:549-567.

Bordovskiy, O. K., 1965, Accumulation of organic matter in bottom sediments: Marine Geol. 3:33-82.

Bray, E. E., and Evans, E. D., 1961, Distribution of *n*-paraffins as a clue to recognition of source beds: Geochim. Cosmochim. Acta, 22:2-15.

Bray, E. E., and Foster, W. R., 1980, A process for primary migration of petroleum: Bull. Am. Assoc. Petr. Geol., 64:107-114.

Brooks, J. D., 1970, The use of coals as indicators of the occurrence of oil and gas: APEA Journal, 10:35-40.

Brooks, J. D., Hesp, W. R., and Rigby, D., 1971, The natural conversion of oil to gas in sediments in the Cooper Basin: APEA Journal, 11:121-125.

Burst, J. F., 1969, Diagenesis of Gulf Coast clayey sediments and its possible relation to petroleum migration: Bull. Am. Assoc. Petr. Geol., 53:73-93.

Casagrande, D. J., and Hodgson, G. W., 1974, Generation of homologous porphyrins under simulated geochemical conditions: Geochim. Cosmochim. Acta, 38:1745-1758.

Claypool, G. E., Love, A. H., and Maughan, E. K., 1978, Organic geochemistry, incipient metamorphism, and oil generation in black shale member of Phosphoria Formation, Western Interior United States, Bull. Am. Assoc. Petr. Geol., 62:98-120.

Claypool, G. E., Lubeck, C. M., Baysinger, J. P., and Ging, T. G., 1977, Organic geochemistry, *in* P. A. Scholle, ed., Geological Studies on the COST No. B-2 well, U. S. Mid-Atlantic Outer Continental Shelf Area: Geological Survey Circular 750, pp. 46-62.

Connan, J., 1974, Time-temperature relation in oil genesis: Bull. Am. Assoc. Petr. Geol. 58:2516-2521.

Connan, J., and Cassou, A. M., 1980, Properties of gases and petroleum liquids derived from terrestrial kerogen at various maturation levels: Geochim. Cosmochim. Acta, 44:1-23.

Cook, E. W., 1974, Green River Shale—oil yields: correlation with elemental analysis: Fuel, 53:16-20.

Cooper, J. E., and Bray, E. E., 1963, A postulated role of fatty acids in petroleum formation: Geochim. Cosmochim. Acta, 27:1113–1127.

Cordell, R. J., 1972, Depth of oil origin and primary migration, a review and critique: Bull. Am. Assoc. Petr. Geol., 56:2029–2067.

———, 1973, Colloidal soap as proposed primary migration medium for hydrocarbons: Bull. Am. Assoc. Petr. Geol., 57:1618–1643.

Demaison, G. J., and Moore, G. T., 1980, Anoxic environments and oil source bed genesis: Org. Geochem., 2:9–31.

Dembicki, H., Meinschein, W. G., and Hattin, D. E., 1976, Possible ecological and environmental significance of the predominance of even-carbon number C_{20}–C_{30} n-alkanes: Geochim. Cosmochim. Acta, 40:203–208.

Dickey, P. A., 1975, Possible migration of oil from source rock in oil phase: Bull. Am. Assoc. Petr. Geol., 59:337–345.

Didyk, B. M., Alturki, Y. I. A., Pillinger, C. T., and Eglinton, G., 1975, Petroporphyrins as indicators of geothermal maturation: Nature (London), 256:563–565.

Didyk, B. M., Simoneit, B. R. T., Brassel, S. C., and Eglinton, G., 1978, Organic geochemical indicators of palaeoenvironmental conditions of sedimentation: Nature (London), 272:216–222.

Dormans, H. N. M., Huntjens, F. H., and van Krevelen, D. W., 1957, Chemical structure and properties of coal. XX. Composition of the individual macerals (vitrinites, fusinites, micrinites, and exinites): Fuel, 36:321–339.

Dow, W. G., 1977, Kerogen studies and geological interpretations: J. Geochem. Expl., 7:79–99.

Emery, K. O., and Rittenberg, S. C., 1952, Early diagenesis of California Basin sediments in relation to origin of oil: Bull. Am. Assoc. Petr. Geol., 36:735–806.

Espitalié, J., Laporte, J. L., Madec, M., Marquis, F., Leplat, P., Poulet, J., and Boutefeu, A., 1977, Rapid method for source rock characterization and for evaluating their petroleum potential and their degree of evolution: Institute Francais du Pétrole and Labofina S.A., March 1977.

Evans, C. R., and Staplin, F. L., 1971, Regional facies of organic metamorphism in geochemical exploration, *in* 3rd Internat. Geochemical Exploration Symposium, Proc. Canadian Institute of Mining and Metallurgy, Special vol. II, Montreal: Canadian Institute of Mining and Metallurgy, pp. 517–520.

Fuex, A. N., 1977, The use of stable carbon isotopes in hydrocarbon exploration: J. Geochem. Expl., 7:155–188.

Galimov, E. M., 1973, Carbon Isotopes in Oil and Gas Geology (in Russian): Moscow, Nedra. English translation: Washington (NASA TT F-G82).

Garrels, R. M., and Perry, E. A. Jr., 1974, Cycling of carbon, sulfur, and oxygen through geologic time, *in* E. D. Goldberg, ed., The Sea, vol. 5: New York, Wiley, pp. 303–336.

Golitsyn, M. S., 1973, The length of the process of coal metamorphism (in Russian): Izv. Akad. Nauk SSSR, Seriya Geologicheskaya, No. 8, 90–97.

Gulyaeva, N. D., Aref'ev, O. A., and Petrov, Al. A., 1978, Pentacyclic hydrocarbons C_{27}–C_{31} in lignites (in Russian), *in* N. B. Vassoevich, ed., Accumulation and Transformation of Organic Material of Recent and Ancient Sediments; Moscow, Nauka, pp. 158–162.

Haseldonckx, P., 1979, Relation of palynomorph colour and sedimentary organic matter to thermal maturation and hydrocarbon generating potential: U.N. ESCAP/CCOP Tech. Publ. No. 6, pp. 41–53.

Héroux, Y., Chagnou, A., and Bertrand, R., 1979, Compilation and correlation of major thermal maturation indicators: Bull. Am. Assoc. Petr. Geol., 63:2128–2144.

Hill, G. A., 1959, Quoted *in* G. Philippi, 1974, Depth of oil origin and primary migration—a review and critique: discussion: Bull. Am. Assoc. Petr. Geol., 58:149–150.

Hitchon, B., 1974, Occurrence of natural gas hydrates in sedimentary basins, *in* I. R. Kaplan, ed., Natural Gases in Marine Sediments: New York, Plenum, pp. 195–225.

Ho, T. T. Y., 1979, Geological and geochemical factors controlling electron spin resonance signals in kerogen: U.N. ESCAP/CCOP Tech. Publ. No. 6, pp. 54–80.

Hobson, G. D., 1961, Problems associated with the migration of oil in "solution": J. Inst. Petr., 47:170–173.

———, 1973, The occurrence and origin of oil and gas, *in* G. D. Hobson, ed., Modern Petroleum Technology, 4th ed.: London, Applied Science Publ., on behalf of Inst. Petr. Great Brit., pp. 1–25.

Hunt, J. M., 1963, Composition and origin of the Uinta Basin bitumens, *in* A. L. Crawford, ed., Oil and Gas Possibilities of Utah, Re-evaluated: Utah Geological and Mineralogical Society, Bulletin 54, pp. 249–273.

———, 1972, Distribution of carbon in crust of Earth: Bull. Am. Assoc. Petr. Geol., 56:2273–2277.

———, 1979, Petroleum Geology and Geochemistry: San Francisco, Freeman.

Hunt, J. M., and Whelan, J. K., 1978, Dissolved gases in Black Sea sediments, *in* D. A. Ross, Y. P. Neprochnov, et al., Initial Reports of the Deep Sea Drilling Project, Volume 42: Washington, U. S. Government Printing Office, pp. 661–665.

Hutton, A. C., Kantsler, A. J., Cook, A. C., and McKirdy, D. M., 1980, Organic matter in oil shales: APEA Journal, 20:44–67.

Jüntgen, H., and Klein, J., 1975, Formation of natural gas from coaly sediments (in German): Erdöl und Kohle, Erdgas, Petrochemie, 28:65–73.

Jurg, J. W., and Eisma, E., 1964, Petroleum hydrocarbons: generation from fatty acid: Science, 144:1451.

Kidwell, A. L., and Hunt, J. M., 1958, Migration of oil in Recent sediments of Pedernales, Venezuela, *in* L. G. Weeks, ed., Habitat of Oil: Symposium Am. Assoc. Petr. Geol., Tulsa, pp. 790–817.

King, L. H., Goodspeed, F. E., and Montgomery, D. S., 1963, Study of sedimented organic matter and its natural derivatives: Mines Branch Report R114, Dept. Mines Techn. Surv., Ottawa.

Kinghorn, R. R. F., and Rahman, M., 1980, The density separation of different maceral groups of organic matter dispersed in sedimentary rocks: J. Petr. Geol., 2:449–454.

Koons, C. B., Bond, J. G., and Peirce, F. L., 1974, Effects of depositional environment and postdepositional history on chemical composition of Lower Tuscaloosa oils: Bull. Am. Assoc. Petr. Geol., 58:1272–1280.

Kvenvolden, K. A., 1967, Normal fatty acids in sediments: J. Am. Oil Chem. Soc., 44:628–636.

Kvenvolden, K. A., and McMenamin, M. A., 1980, Hydrates of natural gas: a review of their geologic occurrence: U.S. Geol. Serv. Cir. 825.

Levorsen, A. I., 1954, Geology of Petroleum: San Francisco, Freeman.

———, 1967, Geology of Petroleum, 2nd ed.: San Francisco, Freeman.

Lewan, M. D., Winters, J. C., and McDonald, J. H., 1979, Generation of oil-like pyrolyzates from organic-rich shales: Science, 203:897–899.

Lijmbach, G. W. M., 1975, On the origin of petroleum: Proc. 9th World Pet. Cong., 2:357–369.

Lopatin, N. V., 1971, Temperature and geologic time as factors in coalification (in Russian): Izv. Akad. Nauk SSSR, Seriya geologicheskaya, No. 3, 95–106.

MacMillan, L., 1980, Oil and gas of Colorado: a conceptual view, *in* H. C. Kent and K. W. Porter, eds., Colorado Geology: Denver, Rocky Mountain Association of Geologists, pp. 191–197.

Magara, K., 1974, Aquathermal fluid migration: Bull. Am. Assoc. Petr. Geol., 58:2513–2526.

Manheim, F., Rowe, G. T., and Jipa, D., 1975, Marine phosphorite formation off Peru: J. Sed. Pet., 45:243–251.

Martin, R. L., Winters, J. C., and Williams, J. A., 1963, Distribution of *n*-paraffins in crude oils and their implications to the origin of petroleum: Nature (London), 199:1190–1193.

Mathews, R. T., Burns, B. J., and Johns, R. B., 1970, Comparison of hydrocarbon distributions in crude oils and shales from Moonie Field, Queensland, Australia: Bull. Am. Assoc. Petr. Geol., 54:428–438.

———, 1971, An approach to identification of source rocks: APEA Journal, 11, part 1, 115–120.

McAuliffe, C., 1966, Solubility in water of paraffin, cycloparaffin, olefin, acetylene, cycloolefin and aromatic hydrocarbons: J. Phys. Chem., 70:1267–1275.

McCarthy, E. D., and Calvin, M., 1967, The isolation and identification of the C_{17} saturated isoprenoid hydrocarbon 2,6,10-trimethyltetradecane from a Devonian shale: Tetrahedron, 23:2609–2619.

McCulloh, T. H., 1969, Geologic characteristics of the Dos Cuadras offshore oil field: U. S. Geol. Surv. Prof. Pap. 679-C, pp. 29–46.

McIver, R. D., 1974, Hydrocarbon gas (methane) in canned Deep Sea Drilling Project core samples, *in* I. R. Kaplan, ed., Natural Gases in Marine Sediments: New York, Plenum, pp. 63–69.

———, 1975, Hydrocarbon occurrences from Joides Deep Sea Drilling Projects: Proc. 9th World Pet. Cong., Tokyo: London, Applied Science Publ., Vol. II, pp. 269–280.

Mechalas, B. J., Meyers, T. J., and Kolpack, R. L., 1973, Microbial decomposition patterns using crude oil, *in* D. G. Ahearn and S. P. Meyers, eds., The Microbial Degradation of Oil Pollutants: Workshop, Center for Wetland Resources, LSU-SG-73-01, pp. 67–79.

Mills, R. V. A., 1923, Natural gas as a factor in oil migration and accumulation in the vicinity of faults: Bull. Am. Assoc. Petr. Geol., 7:14–24.

Milner, C. W. D., Rogers, M. A., and Evans, C. R., 1977, Petroleum transformations in reservoirs: J. Geochem. Expl., 7:101–153.

Momper, J. A., 1978, Oil migration limitations suggested by geological and geochemical considerations, *in* Physical and Chemical Controls on Petroleum Migration, AAPG Continuing Education Course Note Series, No. 8: Tulsa, AAPG, pp. B1–B60.

Neruchev, S. G., and Parparova, G. M., 1972, The role of geologic time in the process of the metamorphosis of coals and dispersed organic matter in sedimentary rocks (in Russian): Akad. Nauk SSSR, Sib. Otdel., Geologiya i Geofizika, No. 10, pp. 3–10.

Neruchev, S. G., Uspenskiy, V. A., Zelichenko, I. A., and Shaks, I. A., 1972, Bituminoid components of the basic genetic classes of dispersed organic material in rocks and their generation during the process of catagenesis: Trudy, Vsesoyuzniy neftyanoy nauchno-issledovatel'skiy geologorazvedochniy Institut (Leningrad), 310:32–57.

Neruchev, S. G., Zelichenko, I. A., Rogozina, E. A., Filatov, S. S., Polovnikova, I. A., and Klimova, L. I., 1978, Characteristics of the catagenetic transformation of humic varieties of dispersed organic material (in Russian), *in* N. B. Vassoevich, ed., Accumulation and Transformation of Organic Material of Recent and Ancient Sediments: Moscow, Nauka, pp. 35–40.

Peake, E., and Hodgson, G. W., 1966, Alkanes in aqueous systems, I. Exploratory investigations on the accomodation of C_{20}–C_{33} *n*-alkanes in distilled water and occurrence in natural water systems: J. Am. Oil Chem. Soc., 43:215–222.

Perry, J. J., and Cerniglia, C. E., 1973, Degradation of petroleum by filamentous fungi, *in* D. G. Ahearn and S. P. Meyers, eds., The Microbial Degradation of Oil Pollutants: Workshop, Center for Wetland Resources, LSU-SG-73-01, pp. 89–94.

Philippi, G. T., 1965, On the depth, time, and mechanism of petroleum generation: Geochim. Cosmochim. Acta, 29:1021–1049.

_____, 1974, Depth of oil origin and primary migration: a review and critique: discussion: Bull. Am. Assoc. Petr. Geol. 58:149–150.

_____, 1975, The deep subsurface temperature controlled origin of the gaseous and gasoline-range hydrocarbons of petroleum: Geochim. Cosmochim. Acta, 39:1353–1373.

Philp, R. P., and Gilbert, T. D., 1980, Application of computerized gas chromatography–mass spectrometry to oil exploration in Australia: APEA Journal, 20:221–228.

Piper, D. Z., and Codispoti, L. A., 1975, Marine phosphorite deposits and the nitrogen cycle: Science, 188:15–18.

Powell, T. G., Cook, P. J., and McKirdy, D. M., 1975, Organic geochemistry of phosphorites: relevance to petroleum genesis: Bull. Am. Assoc. Petr. Geol., 59:618–632.

Powell, T. G., and McKirdy, D. M., 1975, Crude oil composition in Australia and Papua-New Guinea: Bull. Am. Assoc. Petr. Geol., 59:1176–1197.

Powers, M. C., 1967, Fluid-release mechanisms in compacting marine mudrocks and their importance in oil exploration: Bull. Am. Assoc. Petr. Geol., 51:1240–1254.

Price, L. C., 1973, The solubility of hydrocarbons and petroleum in water as applied to the primary migration of petroleum, Ph.D. thesis, University of California, Riverside.

_____, 1976, Aqueous solubility of petroleum as applied to its origin and primary migration: Bull. Am. Assoc. Petr. Geol., 60:213–244.

Pryakhina, Yu. A., 1973, Organic material of the carbonate rocks of the Upper Cretaceous of the northeastern Pre-Caucasus region (in Russian), *in* N. B. Vassoevich, ed., Nature of Organic Material: Moscow, Nauka.

Pusey, W. C., 1973, The ESR-kerogen method... how to evaluate potential gas and oil source rocks: World Oil (April), pp. 71–75.

Robert, P., 1979, Classification of organic material by means of fluorescence and its application to hydrocarbon source rocks (in French): SNEAP Lab. Geol. Boussens, F-31360, Saint-Martory.

Rogers, M. A., McAlary, J. D., and Bailey, N. J. L., 1974, Significance of reservoir bitumens to thermal-maturation studies, Western Canada Basin: Bull. Am. Assoc. Petr. Geol., 58:1806–1824.

Sackett, W. M., 1977, Use of hydrocarbon sniffing in offshore exploration: J. Geochem. Explor., 7:243–254.

Schwebel, D. A., Devine, S. B., and Riley, M., 1980, Source, maturity, and gas composition in the southern Cooper Basin: APEA Journal, 20:191–120.

Seifert, W. K., and Howells, W. G., 1969, Interfacially active acids in a California crude oil: Anal. Chem., 41:554–562.

Seifert, W. K., and Moldowan, J. M., 1978, Applications of steranes, terpanes, and monoaromatics to the maturation, migration, and source of crude oils: Geochim. Cosmochim. Acta, 42:77–95.

_____, 1979, The effect of biodegradation on steranes and terpanes in crude oils: Geochim. Cosmochim. Acta, 43:111–126.

Seifert, W. K., Moldowan, J. M., and Jones, R. W., 1980, Applications of biological marker chemistry to petroleum exploration: Proc. 10th World Petr. Cong., 2:425–438.

Shibaoka, M., Bennett, A. J. R., and Gould, K. W., 1973, Diagenesis of organic matter and occurrence of hydrocarbons in some Australian sedimentary basins: APEA Journal, 13:73–80.

Shimoyama, A., and Johns, W. D., 1971, Catalytic conversion of fatty acids to petroleum-like paraffins and their maturation: Nature (Phys. Sci.), 232:140–144.

_____, 1972, Formation of alkanes from fatty acids in the presence of $CaCO_3$: Geochim. Cosmochim. Acta, 36:87–91.

Simoneit, B. R. T., Mazurek, M. M., Brenner, S., Crisp, P. T., and Kaplan, I. R., 1979, Organic geochemistry of recent sediments from Guaymas Basin, Gulf of California: Deep Sea Res., 26A:879–891.

Smith, B. N., 1975, Carbon and hydrogen isotopes of sucrose from various sources: Naturwiss., 62:390.

Smith, B. N., and Epstein, S., 1971, Two categories of $^{13}C/^{12}C$ for higher plants: Plant Physiol., 47:380–384.

Snarsky, A. N., 1962, Primary migration of oil (in German): Freiberger Forschungsh., C123:63–73.

Snowden, L. R., 1979, Errors in extrapolation of experimental kinetic parameters to organic systems: Bull. Am. Assoc. Petr. Geol., 63:1128–1138.

Spackman, W., Davis, A., and Mitchell, G. D., 1976, The fluorescence of liptinite macerals: Brigham Young University Geology Studies, 22:59–75.

Stahl, W., 1974, $^{13}C/^{12}C$ ratio of North German natural gases (in German): Erdöl und Kohle, 27:623.

_____,1975. Carbon isotope ratios of natural gases (in German): Erdöl und Kohle, 28:188–191.

———, Source rock-crude oil correlation by isotopic type-curves: Geochim. Cosmochim. Acta, 42:1573–1577.
Staplin, F. L., 1969, Sedimentary organic matter, organic metamorphism, and oil and gas occurrence: Bull. Can. Petr. Geol., 17:47–66.
Stutzer, O., 1940, Geology of Coal. (trans. A. C. Noe): Chicago, University of Chicago Press.
Teichmüller M., 1971, Application of coal petrographic methods in oil and gas prospecting (in German): Erdöl und Kohle, 24:69–76.
———, 1975, cited by M-Th. Mackowsky: Microsc. Acta, 77:114.
Teichmüller, M., and Wolf, M., 1977, Application of fluorescence microscopy in coal petrology and oil exploration: J. Microscopy 109:49–73.
Ting, F. T. C., 1975, Fluorescence characteristics of thermal-altered exinites (sporinites): Fuel, 54:200–205.
———, 1979, Source material, geologic age, thermal maturation, and properties of kerogen: AASP 12th annual meeting, Dallas, Oct. 31–Nov. 3.
Tissot, B., 1969, Initial data on the mechanisms and kinetics of petroleum formation in sediments: computer simulation of a reaction scheme (in French): Inst. Francais Pétrole Rev., 24:470–501.
Tissot, B., Califet-Debyser, Y., Deroo, G., and Oudin, J. L., 1971, Origin and evolution of hydrocarbons in Early Toarcian shales: Bull. Am. Assoc. Petr. Geol., 55:2177–2193.
Tissot, B., Deroo, G., and Herbin, J. P., 1979, Organic matter in Cretaceous sediments of the North Atlantic: contribution to sedimentology and paleogeography, *in* M. Talwani, W. Hay, and W. B. F. Ryan, Deep Drilling Results in the Atlantic Ocean: Continental Margins and Paleoenvironment: Washington, AGU, pp. 362–374.
Tissot, B., Durand, B. Espitalié, J., and Combaz, A., 1974, Influence of nature and diagenesis of organic matter in formation of petroleum: Bull. Am. Assoc. Petr. Geol., 58:499–506.
Tissot, B., and Espitalié, J., 1975. Thermal evolution of organic matter in sediments: application of a mathematical simulation (in French): Inst. Francais Pétrole Rev., 30:743–777.
Tissot, B. and Pelet, R., 1971, New data on the mechanism of formation and migration of petroleum: mathematical simulation and application to prospecting (in French): Proc. 8th World Pet. Cong. 2:35–46.
Tissot, B., Pelet, R., Roucache, J., and Combaz, A., 1977, Use of alkanes as geochemical fossil indicators of geological environments (in French): *in* R. Campos and J. Goni, eds., Advances in Organic Geochemistry 1975: Madrid, Enadisma, pp. 117–154.
Tissot, B. P., and Welte, D. H., 1978, Petroleum formation and occurrence: New York, Springer-Verlag.
Tucholke, B. E., Bryan, G. M., and Ewing, J. I., 1977, Gas-hydrate horizons detected in seismic-profile data from the western North Atlantic: Bull. Am. Assoc. Petr. Geol., 61:698–707.
Vandenbroucke, M., Albrecht, P., and Durand, B., 1976, Geochemical studies on the organic matter from the Douala Basin (Cameroon)—III. Comparison with the Early Toarcian Shales, Paris Basin, France: Geochim. Cosmochim. Acta, 40:1241–1249.
Waples. D. W., 1976, Time-temperature relation in oil genesis: discussion: Bull. Am. Assoc. Petr. Geol., 60:884–885.
———, 1977, C/N ratios in source rock studies: Min. Ind. Bull. Colo. Sch. of Mines, 20, No. 5.
———, 1979, Simple method for source rock evaluation: Bull. Am. Assoc. Petr. Geol., 63:239–245.
———, 1980, Time and temperature in petroleum exploration: application of Lopatin's method to petroleum exploration: Bull. Am. Assoc. Petr. Geol., 64:916–926.
Waples, D. W., and Sloan, J. R., 1979, Carbon and nitrogen profiles in deep sea sediments; new evidence for bacterial diagenesis at great depths of burial, *in* G. deV. Klein, K. Kobayashi, et al., Initial Reports of the Deep Sea Drilling Project, Vol. 58: Washington, U. S. Government Printing Office, pp. 745–754.
Welte, D. H., 1965, Relation between petroleum and source rock: Bull. Am. Assoc. Petr. Geol., 49:2246–2267.
———, Petroleum exploration and organic geochemistry: J. Geochem. Expl., 1:117–136.
Welte, D. H., Hagemann, H. W., Hollerbach, A., Leythaeuser, D., and Stahl, W., 1975a, Correlation between petroleum and source rock: Proc. 9th World Pet. Cong., 2:179–191.
Welte, D. H., Kalkreuth, W., and Hoefs, J., 1975, Age-trend in carbon isotopic composition in Paleozoic sediments: Naturwiss., 62:482–483.
Welte, D. H., and Waples, D. W., 1973. Even *n*-alkane predominances in sedimentary rock extracts (in German): Naturwiss., 60:516–517.
White, D., and Thiessen, R., 1913, The origin of coal: Bur. Mines Bull., No. 38.
White, R. S., 1979, Gas hydrate layers trapping free gas in the Gulf of Oman: Earth Plan. Sci. Lett., 42:114–120.
Williams, J. A., 1974, Characterization of oil types in the Williston Basin: Bull. Am. Assoc. Petr. Geol., 58:1243–1252.
Yen, T. F., 1972, Present status of the structure of petroleum heavy ends and its significance to various technical applications: Am. Chem. Soc. Meeting Preprints, F:102–114.
Zieglar, D. L., and Spotts, J. H., 1978, Reservoir and source bed history of Great Valley, California: Bull. Am. Assoc. Petr. Geol., 62:813–826.

INDEX

A

Abyssal sediments, 17, 19
Accumulation. See Petroleum, accumulation of
Activation energy, 31, 33, 95, 96, 100
Aggregates of asphaltenes. See Asphaltenes, aggregated
Alanine, 10
Algae, 1, 11, 13, 15–18, 87. See also Photoplankton; Plankton
    blue-green, 87. See also Phytoplankton
Algal organic material. See Kerogen type, algal
Alganite, 22, 23, 44, 69, 70, 76
Alkanes, 5, 7, 27. See also Hydrocarbons, saturated; n-Alkanes
Alkenes, 6, 7
Alumina, in column chromatography, 60
Amino acids, 10, 13, 14
Ammonia, release during diagenesis, 36, 43
Amorphous organic material. See Kerogen type, amorphous
Anaerobes. See Bacteria, anaerobic
Anaerobic decomposition. See Bacterial decomposition, anaerobic
Analytical methods, 58–65
Anoxic sediments, 15, 18, 19
API gravity, changes in, 89, 91
    application in oil–oil correlations, 92, 93, 119–125
Aromatic sheets, 24, 43, 45
Aromatic compounds, 7, 10, 24, 27. See also Hydrocarbons, aromatic
Aromaticity, 21, 85
Aromatization, 27, 39, 43, 45, 84, 85
Arrhenius equation, 95
Arrhenius plot, 95
Asphaltenes, 23, 25, 30 35, 36, 39, 60
    aggregated, 24
    ^{13}C in, 91
    changes in, 88, 89, 91
    migration of, 52, 88, 91
    precipitation of, 25, 27, 60, 89–91
    structure of, 24
Athabasca tar sands, 45, 57
Australia, 15, 93

B

Bacteria, 11, 13, 38, 87
    aerobic, 14, 16, 17, 89
    anaerobic, 14, 16, 17, 21, 26, 89
    contribution to kerogen, 22
    methane production by, 21, 30, 83
Bacterial decomposition, 11–19, 26. See also Petroleum transformation, biodegradation
Bacteriocides, 10, 15, 19
Baltimore Canyon, 77
Baltimore Dome, 33
Bark, 14
Benzene, 7, 26, 85
    solubility of, 49
    as solvent, 60
    benzene-methanol as extraction mixture, 59
BFOC, 67, 69, 86
    application in source-rock evaluation, 107, 108, 117, 118. See also TOC, application in source-rock evaluation
Biodegradation. See Petroleum transformation, biodegradation
Biological productivity. See Productivity, biological
Biopolymers, 13, 14. See also Polymers
Bisnorhopane, 7, 8
Bitumen, 2, 20
    adsorption of, 53, 54, 74
    analysis of, 36–42, 46, 59–62
    application in correlations, 3, 42

catagenetic, 32–36, 39, 64
    formation of, 31–36, 43, 46, 86
composition of, 20, 23–30, 33, 36–42
crackin of, 34, 35, 84–86, 94
diagenetic, 34–36, 39, 64
    migration of, 53, 82
elemental analysis of, 36
expulsion of, 54
    efficiency of, 66, 68, 74, 75, 83
extraction of, 59, 60, 63
heteroatoms in, 30, 36
inherited. *See* Bitumen, diagenetic
as maturity indicator, 34–42, 74
migration of, 20, 26, 30, 34, 35. *See also* Bitumen, diagenetic, migration of; Bitumen, expulsion of; Migration
solidifed, 21, 30, 52
as contaminant, 80
Bitumen content
    application in source-rock evaluation, 35, 36, 67, 107, 108, 113–115, 117, 118
    spurious, 36, 59, 108, 115
Bitumen-free organic carbon. *See* BFOC
Black Sea, *17*, 18
Branching, in organic compounds, 5
Buoyancy. *See* droplets, buoyancy of
Butane, *6*

C

C_3 pathway. *See* Calvin cycle; Plants, C_3
C_4 pathway. *See* Dicarboxylic acid metabolic pathway; Plants, C_4
$\delta^{13}C$. *See* Isotopes, carbon
C_{15+} bitumen, 35, 60. *See also* Bitumen
California, 47, 53, 56, 106
Calvin cycle, 87
Canada, 47
Canning of samples. *See* Samples, canning of
Cap rock. *See* Seals, reservoir
Capillary entry pressure. *See* Entry pressure
Carbohydrates, 10, 15, 87
Carbon, organic. *See* BFOC; TOC
Carbon cycle, 11–13
Carbon dioxide
    in carbon cycle, 11–13, 87
    diagenetic, 11, *12*, 30, 36, 43
    in isotope analyses, 65
    isotopic composition of, *88*
    in organic carbon analyses, 59, 63
    in Rock-Eval analyses, 63, 64
Carbon isotopes. *See* Isotopes, carbon
Carbon Preference Index. *See* CPI
Carbonate content, measurement of, 59
Carbonates
    catalysis by, 33
    expulsion of bitumen from, 75
    isotope ratios of, *88*
    organic networks in, 52
    recrystallization of, 52
    removal by acid, 59, 62
    sulfur in, 26, 31
    thermal decomposition of, 63
    TOC measurements in, 59
Carboxyl groups, 8, 49, *50*
Carboxylic acids, *See* fatty acids
β-carotane. *See* perhydro-β-carotene
β-carotene, 29
Catagenesis. *See* Kerogen, catagenesis of
Catalysis, 33, 95. *See also* Carbonates, catalysis by; Clays, catalysis by
Cellulose, 10, 13, 15, 21, 23
Chelation, 29
Chips, drilling. *See* Cuttings
Chitin, 13
Chloroform, as extraction solvent, 59
Chlorophyll, *8*, 9, 28
Cholestane, 7, *8*, 87, 130
Cholesterol, *8*
Chromatography. *See* Column chromatography; Gas chromatography
Cis configuration, 42
Clathrates. *See* Gas hydrates
Clays
    adsorption by, 24. *See also* Migration, factors affecting, adsorption
    catalysis by, 33
Coal, 2, 11–16, 43–45, 75
    $\delta^{13}C$ of, 65, *88*
    as kerogen, 2, 14, 20, 21
Coal measures, 15
Coal petrology, 22, 45
Coal rank, 43–45
Coal swamps, 14–16
Color of organic material. *See* TAI
Column chromatography, 60
Computer, attached to gc-ms, 61
Condensates, 21, 31
Conduits for migration. *See* Migration, conduits for
Contamination, 36, 58, 59, 73, 74, 80, 108, 115, 118
Continuous phase migration. *See* Migration, mechanisms of, continuous phase
Cocking time, in kerogen catagenesis. *See* Petroleum formation, time and temperature in
Correlations. *See also* Fingerprinting
    gas, 94
    negative evidence in, 92
    oil–bitumen. *See* Correlations, oil–source rock
    oil–oil, 3, 88–93
        examples of, 91–93, 119–125
    oil–source rock, 3, 88–91, 93
        examples of, 127–131
C.O.S.T. B-2 well, 77, 78
C.O.S.T. #1 well, 78–80, *103*, 104
CPI, 37
    application in correlations, *89*, 91, 119, 123, 125, 129
    application in environmental interpretations, 42, 131
    as maturity indicator, 37–39, 42
Cracking. *See* Bitumen, cracking of; Petroleum transformation, cracking

Cresol, *10*
Crude oil. *See* Petroleum
Cuttings, 58, 59, 118
Cyclization, 43, 84, 85
Cyclo-. *See* Hydrocarbons, cyclic
Cycloalkanes. *See* Hydrocarbons, cyclic
Cyclohexane, 4–6, 21, *23*, *26*, 29
Cyclohexene, *6*
Cyclopentane, *6*

D

Deasphalting. *See* Asphaltenes, precipitation of
Decay. *See* Bacterial decomposition
Decomposition. *See* Bacterial decomposition; Bitumen, cracking of; Kerogen, catagenesis of; Petroleum transformation, cracking
Deep Sea Drilling Project, *17*, 18, 58
Degradation. *See* Bacterial decomposition; Petroleum transformation, biodegradation
Delocalization of electrons, 43, 85. *See also* Free radicals
Deltas, 15, *16*, 18, 47
Depositional environments, 16–19, 86, 87
Diagenesis, 1, 13–19, 22, 30, 34–36, 38, 44, *53*
Dicarboxylic acid metabolic pathway, 87
Diesel fuel, as contaminant, 59, 108
Diffusion. *See* Migration, mechanisms of, diffusion
Diploptene, *8*
Disproportionation, 27, 88, 91
Double bonds, 6, 7, 33
Drilling chips. *See* Cuttings
Drilling fluid, contamination by, 36, 59, 108
Droplets
  buoyancy of, 55, 56
  deformation during migration, 51
  role in migration. *See* Migration, mechanisms of, droplets
DSDP. *See* Deep Sea Drilling Project

E

E. *See* Expulsion efficiency factor
Electron spin resonance. *See* Esr
Electrons, delocalization of. *See* Delocalization of electrons
Elemental analysis. *See* Bitumen, elemental analysis of; Kerogen composition, elemental
Energy of activation. *See* Activation energy
Entry pressure, 51, 54
EOM. *See* bitumen
Erosion, effect on geologic reconstructions, *98*, 99, 106
Error
  analytical, 46, 58, 59, 80, 82, 111
  interpretive, 80, 111
Esr, 45, *67*, 86
Ethane, 4–7, 21, 26, 83
Ethene, 6, 7
Ethylene. *See* Ethene
Exinite, 22, *23*, 69, *70*
Expulsion. *See* Bitumen, expulsion of
Expulsion efficiency factor, 75, 83

Extractable organic material
Extraction. *See* Bitumen, extraction of

F

Farnesane, *28*
Fats, *13*, 15, 21. *See also* Fatty acids; Lipids; Wax, plant
Fatty acids, 8, 14, 21, 25, 26, 49, *50*
Faults. *See also* Migration, factors affecting, faults
  effect on geochemical profiles, 114
  effect on geologic reconstructions, 99
  timing of, 57
Fingerprinting, 29, 89, 120, 122, 124, 129. *See also* Correlations
Fixed carbon, 44
Flavanone, *13*
Fluorescence, 45, 46, *67*, 86
Fracturing, 82, 115. *See also* Source rock, as reservoir
Free radicals, 43, 45, 83–85
French Petroleum Institute, 25, 38, 39, 60, *64*
Fulvic acids, 14, 20
Functional groups, 7
Fungi, 58

G

Gas, 15, *16*. *See also* Methane
  biogenic. *See* Methane, biogenic
  composition of, 30, 31, 94
  dissolved in petroleum, 31
  dry, 21, 31, 94
    deadline for, *102*, 105
  isotopic composition of, 31, *88*, 94
  role in migration. *See* Migration, factors affecting, gas
  thermal, 21, 31, *34*, 35, 42, *53*, 54, 57, 83–86, *88*, 94
    from bitumen and petroleum. *See* Bitumen, cracking of; Petroleum transformation, cracking
  wet, 21, 31, 94
    deadline for, *102*, 105
  wetness of, 31, 94
Gas cap, 31
Gas chromatograms, examples of, 61, 62, 121
Gas chromatograph-mass spectrometer, 3, 42, 61, 124
Gas chromatography, 2, 3, 60–64
  application in correlations, 108, 120–124, 129
Gas deasphalting. *See* Asphaltenes, precipitation of
Gas hydrates, 31
Gas-source potential, estimates of, 83–86
Gasoline-range hydrocarbons. *See* hydrocarbons, gasoline-range; Hydrocarbons, light
Gc-ms. *See* Gas chromatograph-mass spectrometer
Generative capacity. *See* Kerogen, hydrocarbon generative capacity of
Geomonomer, 13, 14. *See also* Monomer
Geopolymer, 14, 19, 20. *See also* Fulvic acids; Humic acids; Kerogen; Polymer
Geothermal gradient, 33, 97–99, 101, 106
  relation to lithology, 98, 99
Gilsonite, 53
Globules. *See* Droplets

Glycerol, 13
Graphitization, 43
Green River Formation, *16*, 18, *22*, 23, 52, 68, 75, 76
Growth environments of organisms, *16*, *17*, 66, 67, 86, 87
Growth faults. *See* Migration, factors affecting, faults
Gulf Coast, 16–18, 57, 78–80, *103*, 104

H

H/C ratio. *See* Kerogen composition, H/C
HCl. *See* Carbonates, removal by acid
Heavy oil. *See* Tar
Heptane, as solvent in column chromatography, 60
Heteroatoms, 8
    in asphaltenes, 24, 25, 90
    cleavage of bonds to, 33, 83
    loss during diagenesis, 36, 43
    in refined products, 108
Heterocompounds, 8
    biodegradation of, 89
    solubility of, 49
HF. *See* Silicates, removal by HF
Homologous series, 61
Hopanes, 87. *See also* Bisnorhopane; Triterpanes
Humic acids, 14, 20
Hydrates. *See* Gas hydrates
Hydrocarbon-generative capacity. *See* Kerogen, hydrocarbon generative capacity of
Hydrocarbons, 4–8
    acyclic, 39, *40*
    application in correlations, 119–124
    aromatic, 7, 24, 29, *44*
        in bitumen and petroleum, 23, *25*, 26, 30, 36, *37*, 39, *40*
        separation of, 60
    biodegradation of, 14, 26, 89
    branched, 5, 27, 60–62
    cyclic, 6, 7, 27, *44*, 60–62. *See also* Cyclization; Hydrocarbons, aromatic; Hydrocarbons, polycyclic
    gasoline-range, 42, *89*, 91
    isoprenoid, 7, 28, 29
        application in correlations, 67. *See also* Pristane/phytane ratio
        biodegradation of, 26
    light. *See also* Hydrocarbons, gasoline-range
        application in correlations, 91, 93
        formation of, 33, 42, 91
        as maturity indicators, 42
        solubility of, 49, 50, 89
    naphthenic, *40*
    naphthenoaromatic, 29, *37*, *40*
    polycyclic, 30, 39, 93. *See also* Steranes; Triterpanes
    saturated, 6, 23, 27, 30, *37*, 39, 91
        analyses of, 60–62
        separation by column chromatography, 60
        solubility of, 48, 49, 89
    unsaturated, 6
Hydrochloric acid. *See* Carbonates, removal by acid
Hydrodynamics. *See* Migration, factors affecting, hydrodynamics
Hydrofluoric acid. *See* Silicates, removal by HF
Hydrogen index, 64, 68
Hydrogen sulfide, 30, 31, 36
Hydrogenation, 7, 26
Hydrophilic, 49, *50*
Hydrophobic, 48–50
Hydroxyhopane, 40–42
Hydroxyl groups, 10
Hypersalinity, 87

I

Igneous intrusions. *See* Intrusions, igneous
Inertinite, 22, 23, 44, 69, 70
Inspissation. *See* Petroleum transformation, inspissation
Intrusions, igneous, 33, 100
Iron sulfides, 30. *See also* Marcasite; Pyrite
Isoprenoids. *See* Hydrocarbons, isoprenoid
Isotopes, carbon, 31, 64, 65, 87, *89*
    application in correlations, 67
        examples, 91–93, 119–125, 128–130
    application in gas analyses. *See* Gas, isotopic composition of
    effect of migration on, 128, 131
    as environmental indicators, 67, 87
    measurement of, 65
    relation between kerogen and bitumen, 128, 131
    standards for, 65
    values of, 87, *88*, 91

K

Kerogen, 1, 12, 14, 20
    adsorption of bitumen by, 53, 54
    analysis of, 44–46, 62–65
    catagenesis of, 1, 22, 23, 32–38, 43–46, 83, 84
        kinetics of, 32–36, 73, 95, 96
    formation of, 14–19, 21, 22
    hydrocarbon generative capacity of, 2, 35. *See also* Gas source potential, estimates of; Oil already generated; Total oil
        estimation of, 68–77, 84–86
    hydrogen index of. *See* Hydrogen index
    immature H/C ratios for, 69, *71*, 77, 80, 81, 110
    isolation of, 62
    maturation pathway for, 22, *23*, 69, 70
    metagenesis of, 31, 33, 43, 83, 94
    oxidation of, 63
    oxygen index of. *See* Oxygen index
    purification of, 63
    quality of, 66–71, 86
        application in source-rock evaluations, 77–83, 109–118
    quantity of, 66–69
        application in source-rock evaluations, 77–83, 109–118
    structure of, 21, 22, 42–46
        cross-linking in, *43*
    thermal composition of. *See* Kerogen, catagenesis of

visual analysis of, 22, 44–46, *67*, 69. *See also* TAI
    application in source-rock evaluations, 78, 79, 109–115, 127, 128
Kerogen composition, 21–23, 42–46
    application in source-rock evaluation, 77, 80–82, 109–118
    aromaticity, 21, 43–45
    changes during maturation, 19, 22, 23, 44, 69, 70
    elemental, 22
    H/C ratio of, 22, *23*, 44, 63, 64, 69, 84–86
    nitrogen, 8, 21, 22, 63
    oxygen, 4, 8, 21–23, 63, 64
    sulfur, 21, 22
Kerogen concentrate, 62, 63
Kerogen type, 21–23
    algal, 16–19, 21, 45. *See also* Alginite
    amorphous, 69. *See also* Kerogen, algal
    determination of, 21–23, 69–71
    gas-generative, 68, 86
    oil-generative, 35, 68, 86
    woody, 15–17, 21–23
Kinetics. *See* Kerogen, catagenesis of, kinetics

L

Lacustrine depositional environments, *16*
Lakes, *16*
Leaf wax, 15, 21, 22, 25. *See also* Wax, plant
Leakage. *See* Traps, leakage from
Leco carbon analyzer, 59
Lignin, 10, 13–15, 21
Lipid, 15, 21, 87
Liptinite, 22, 69
Lopatin's method, 3, 96–106
    examples of, 96–106, 131–137
Los Angeles Basin, *35*, 48, 80–82
Lycopane, 28
Lycopene, 29

M

M. *See* Maturity factor
Maceral analysis. *See* Kerogen, visual analysis of
Macerals, organic, 22, 23
Marcasite, 26
Mass spectrometer. *See also* Gas chromatograph-mass spectrometer
    operation of, 61
Mass spectroscopy
    of branched–cyclic hydrocarbons, 61
    in carbon-isotope measurements, 65
    in group-type analyses, 29
    in porphyrin analyses, 62
Maturation pathway. *See* Kerogen, maturation pathway for
Maturity factor, 73–77, 82
    examples of, 79–81, 128
Metagenesis. *See* Kerogen, metagenesis of
Metal in petroleum, 29
Methane, 4–6, *See also* Gas
    biogenic, 21, 30, 83, 94
    H/C ratio of, 84
    thermal. *See* Gas, thermal
    use in deasphalting, 25
Methane hydrates. *See* Gas hydrates
Methanol, as solvent in column chromatography, 60
Micells. *See* Migration, mechanisms of, micells
Microbial transformations. *See* Bacterial decomposition
Microfractures. *See* Migration, mechanisms of, microfracturing
Microorganisms. *See* Algae; Bacteria; Phytoplankton
Microscopic organic analysis. *See* Kerogen, visual analysis of; TAI
Microscopy
    reflected light, 2, 3. *See also* Vitrinite reflectance
    transmitted light. *See* kerogen, visual analysis of; TAI
Middle East, 47. *See also* Saudi Arabia
Migration, 47–57
    compositional changes during, 20, 26, 30, 49, 88, *89*, 91
    examples of, 92, 120, 122, 128, 131
    conduits for, 48, 57
        turbidite channels, 57
    depth of, 53
    distance of, 20, 47, 48
    efficiency of, 18, 57, 77, 80, 82. *See also* Bitumen, expulsion of, efficiency
    factors affecting
        adsorption, 24, 48, 53, 54
        clay dehydration, 52, 53
        compaction water, 52
        CO_2, 55
        faults, 48, 57, 80
        gas, 52, 55
        hydrodynamics, 48, 51, 54–56
        lateral facies changes, 18
        pore diameters, 50–52, 54
        pressure, 51–53
        surfactants, 50
    lateral, 47, 48, 57
    mechanisms of
        continuous phase, 47, 52–54
        diffusion, 51
        droplets, 47–56
        micells, 47–50, 52, 55
        microfracturing, 51–54
        solution, 47–55
    primary, 47–54
    secondary, 47, 48, 54, 55
    speed of, 48
    timing of, 53, 54
    vertical, 48
Migration window, 54
Mississippi delta, 18
MOA. *See* kerogen, visual analysis of
Mold, 58
Molecular sieves, 60
Monomer, 10
Monterey Formation, 47, 53

N

n-Alkanes, 5, *6*
　application in correlations, 91–93, 122, 123, 125, 129–131
　biodegradation of, 26, 89
　carbon preference in, 27, 28, 37. *See also* CPI
　environmental significance of, 25, 28, 37, 38, 86
　as maturity indicators, 28, 37, 38
　separation from saturated hydrocarbons, 60
　solubility of, *50*
n-Alkane distributions, 27, 30, 37–39, 50, *51*, *61*
Naphthalene, 7
Naphthenes, 27
Natural gas. *See* Gas
Nitrogen, organic. *See* Kerogen composition, nitrogen
Nitrogen gas, formation during catagenesis, 36
Nmr, 39
Normal. *See n*-Alkanes
n-Paraffins. *See n*-Alkanes
NSO compounds, 8, 25, 29
Nuclear magnetic resonance spectroscopy. *See* Nmr
Nutrients, essential, 14, 15, 18

O

O/C ratio of kerogen. *See* Kerogen composition, oxygen
Odd-carbon preference, 27, 28, 37. *See also* CPI
Oil. *See* Petroleum
Oil already generated, 66, 67, 76, 77
　profiles, 77–82, 110–118, 127, 128
Oil-generative capacity. *See* Kerogen, hydrocarbon generative capacity of
Oil–oil correlations. *See* Correlations, oil–oil
Oil shales, 75. *See also* Green River Formation
Oil-source capacity. *See* Kerogen, hydrocarbon generative capacity of
Oil–source rock correlations. *See* Correlations, oil–source rock
Oil spill, 56
　application of oil–oil correlations in, 123–125
Olefins, 6
Organic material. *See also* Bitumen; Kerogen; Petroleum
　decomposition of. *See* Bacterial decomposition
　preservation of, 14–19
　reworking of, 23, 46
　terrestrial, 15, *16*, 18, 19, 25, 28
Oxidation
　biological. *See* Bacterial decomposition
　nonbiological, 19, 23, 26
Oxygen in organic matter. *See* Bitumen, composition of; kerogen composition, oxygen; Petroleum composition, heteroatoms
Oxygen
　measurement of in kerogen, 64
　in water and sediments, 14, 15, 19
Oxygen index, 64

P

Paleotemperatures. *See* Geothermal gradient
Palmitic acid, 8
Palynology techniques, application to kerogens, 22, 62, 63
Paraffinicity, *89*
Paris Basin, *22*, *35*, 39, *40*
PDB. *See* Isotopes, carbon, standards for
Peat, 15, *16*
PeeDee Belemnites. *See* Isotopes, carbon, standards for
Pelagic sediments, 18. *See also* Abyssal sediments
Pentane, 5, *6*
　precipitation of asphaltenes by, 60
Perhydro-β-carotene, 28
Permafrost, 31
Permil, in isotope ratios, 65
Petroleum, 20
　accumulation of, 1, 47, 55–57, 68
　analysis of, 59–62, 65
　API gravity of. *See* API gravity
　comparison with bitumen, 30
　as contaminant. *See* Diesel fuel
　correlations of. *See* Correlations, oil–oil; Correlations, oil–source rock
　deadline for, 102
　gas generation from. *See* Petroleum transformation, cracking
　leakage from reservoirs, 56
　migration of. *See* Migration
　organic origin of, 1
Petroleum–bitumen correlations. *See* Correlations, oil–source rock
Petroleum composition, 4, 23–25, 31, *38*, 39
　changes in. *See* Petroleum transformation
　effect of biodegradation on. *See* Petroleum transformation, biodegradation
　heteroatoms, 4, 8, 25, 26, 30
　isotopic, *88*
　waxiness, 25, 27, 93, 131
Petroleum formation. *See also* Kerogen, catagenesis of
　depth of, 50, 53
　hot and deep theory, 49
　kinetics of. *See* Kerogen, catagenesis of, kinetics
　principal zone of, *35*, *53*, 73
　time and temperature in, 3, 33, 78, 95–106
　timing of, 54, 57
Petroleum transformation
　biodegradation, 25, 56, 88–92, 122, 124
　cracking, 21, *53*, 57, 94
　deasphalting. *See* Asphaltenes, precipitation of
　inspissation, 56
　water washing, 26, 89, 91
Petroleum window, 54. *See also* petroleum formation, principal zone of
Phenols, 10
　as bacteriocides, 10, 14, 15, 19
　contribution to woody kerogen, 21
　solubility of, 26
　as surfactants, 49
Phosphatic shales, 18. *See also* Phosphoria Formation
Phosphoria Formation, *17*, 18, 48
Phosphorites, 18
Phosphorus, as nutrient, 7
Photosynthesis, 11, 12, 87

Phytane, 7, *8*, *23*, 87
    application in correlations, 123, 125. *See also* Pristane/phytane ratios
    application in environmental interpretations, 87
Phytol, *8*
Phytoplankton, *12*, 13, 18, 21. *See also* Algae
Pinchouts, 55
Plankton. *See also* Phytoplankton; Zooplankton
    isotopic composition of, 87, *88*
Plants. *See also* Phytoplankton
    C_3, 87, *88*
    C_4, 87, *88*
    marine, $\delta^{13}C$ of, 87, *88*
    terrestrial, 12, 13, 15, *16*, 93
        $\delta^{13}C$ of, 87, *88*
Plastic, as contaminant, 58
Plutons, 33
Polar compounds. *See* Resins, petroleum
Pollen
    contribution to kerogen, 15, 21, 22
    in TAI measurements, 45
Polymers, 10, 20. *See also* Biopolymers; Geopolymers
    kerogen as, 2, 14, 20
Polymerization, 14, 20, 45
Polysaccharides, 10
Pore diameters, 50–52, 54
Pores, plugging by asphaltenes, 27
Porphyrins, 9, 29
    analysis of, 62
    application in correlations, 37, 92, 93, 119–125, 128–131
    changes in
        by migration, *89*, 91
        by refining, 91, 128
    from chlorophyll, 9, 29
    copper, 29
    as maturity indicators, 37
    migration of, 52
    nickel, 9, 29, 62, 91
    Ni/V ratio in, 91
    as resins, 24, 29
    vanadyl, 29, 62, 91
Potential, fluid. *See* Migration, factors affecting, hydrodynamics
Preservation. *See* Organic material, preservation of
Pressure, role in migration. *See* Migration, factors affecting, pressure
Primary migration. *See* Migration, primary
Pristane, 28, *62*. *See also* Pristane/phytane ratios
Pristane/phytane ratios
    application in correlations, 93, 119, 120, 123–125, 129–131
    environmental significance of, 87
Productivity, biological, 16–18
Profiling of wells, 36, 72
    examples, *34*, *35*, *37*, *77–81*, *111*, *115*, *118*, *128*, *129*
Prokaryotic organisms, 87. *See also* Algae, blue-green; bacteria
Propane, *6*
Propene, *6*
Propylene. *See* propene
Proteins, 10, 13, 15

Pyrite. *See also* Iron sulfides
    formation of, in sediments, 26
    removal from kerogen, 63
Pyrobitumen. *See* Bitumen, solidified
Pyrolysis, 46. *See also* Rock-Eval
    application in source-rock evaluations, 67, 70, 71, 76, 111
    examples of, 80–82
    as simulation of catagenesis, 63
PZOF. *See* Petroleum formation, principle zone of

Q

Q_1. *See* Quantity factor
Q_2. *See* Quality factor
Quality factor, 69, 75, 76
Quantity factor, 69, 75, *76*, 82

R

R_O. *See* Vitrinite reflectance
Radicals. *See* Free radicals
Recent sediments, *n*-alkanes in, *38*, *39*
Refined products, composition of, 91, 108
Reflectance. *See* Microscopy, reflected light; Vitrinite reflectance
Reservoir seals. *See* Seals, reservoir
Reservoir transformations of petroleum. *See* Petroleum transformation
Reservoirs
    breached, 56
    fractured shale, 20, 47
    plugging of by asphaltenes, 25, 27
    TTI values of, 104, 105
Resins, petroleum, 24, 25, *35*, 50. *See also* NSO compounds
    compositional variations in, 30, 36, 88, 91
    loss during migration, 50, 88
    separation and isolation of, 60
Resins, tree, 15
Respiration, 11, *12*
Rock-Eval, 46, 60. *See also* Pyrolysis
    application in source-rock evaluations, 67, 68, 86
    operation of, 63, 64
Rubber, as contaminant, 59, 80

S

S_1 peak, 63
S_2 peak, 63, 64
Samples
    analysis of. *See* Analytical methods; Bitumen, analysis of; Kerogen, analysis of
    canning of, 58
    contamination of. *See* Contamination
    screening of, 59
Santa Barbara Channel oil spill, 56
Sapropel, 18
Saudi Arabia, 57
Screening. *See* Samples, screening of
Seals, reservoir, 56
Secondary migration. *See* Migration, secondary
Seeps, petroleum, 56
Serine, 10

Siberia, 83
Sieves, molecular. *See* Molecular sieves
Silica gel, in column chromatography, 60
Silicates, removal by HF, 62
Sitostane, 87
Solvation, 48
Solvents. *See* Bitumen, extraction of
Source rock
    generative capacity of. *See* Kerogen, hydrocarbon generative capacity of
    as reservoir, 20, 82
Source-rock evaluation, 75–77
    examples of, 77–82, 107–118
Source rock–oil correlations. *See* Correlations, oil-source rock
Soxhlet extraction, 59, 60
Spore darkening, *See* TAI
Spores, 21
Sporopollenin, 13
Squalane, 28, *62*
Squalene, 29
Staining, in cores, 127–129. *See also* Contamination
Steranes, 7
    analysis of, 61
    application in correlations, *67*, 91–93
        examples, 124, 129–131
    biodegradation of, 89
    as environmental indicators, 29, *67*, 87
    as maturity indicators, 39, *40*
Steroids, 40, 87
Stigmastane, 87
Stratification of water bodies, 16–18
Structure development, timing of, 57
Sugars, 10, *13*, 14
Sulfate, bacterial reduction of, 26
Sulfides. *See* Iron sulfides; Marcasite; Pyrite
Sulfur. *See also* Bitumen, heteroatoms in; Kerogen composition, sulfur; Petroleum composition, heteroatoms
    elemental, 26, 102
    organic, application in correlations, 119–125
Surfactants, 50
Swamps. *See* Coal swamps

T

TAI, 45
    application in source-rock evaluations, *67*, 82
        examples, 107–118, 127, 128
    correlation with R_O, 72, *102*
    correlation with TTI, *102*
    limits of usefulness, 45, 46, 36
    relation to maturity factor, 74
    uncertainty in, 72
Tannins, 14
Tar, reservoir, 27, 56
Tar sands. *See* Athabasca tar sands
Temperature
    paleo-. *See* Geothermal gradient
    role in petroleum formation. *See* Petroleum formation, time and temperature in

Temperature–time relationship. *See* Petroleum formation, time and temperature in
Terrestrial organic material. *See* Organic material, terrestrial
Texas, *38*, 78–80, *103*, 104
Thermal Alteration Index. *See* TAI
Time, role in petroleum formation. *See* Petroleum formation, time and temperature in
Time-Temperature Index of Maturity. *See* TTI
Tissot diagram, 22, *23*, 64, 69
TOC, 20
    application in source-rock evaluations, 67, 69
        examples, 77–82, 109–115, 127, 128
    measurement of, 59
Toluene, 7
Total gas. *See* Gas-source potential, estimates of
Total oil, 66–69, 75–77
    application in source-rock evaluations, 77–82, 110–118, 128, 129
Total organic carbon. *See* TOC
Trans configuration, 40, 42
Transformation ratio, 64
Transport of sedimentary material, 14–17
Trapping mechanisms, 55, 56
Traps. *See also* Reservoirs
    leakage from, 56
    relation to generation sites, 48, 57
Trees, 15
Triterpanes, 7
    applciation in correlations, *67*, 91
        examples, 92, 124, 129–131
    biodegradation of, 89
    as environmental indicators, 29 *67*, 87
    geochemical transformations of, 40–42
Triterpenoids, 40, *41*, 87
TTI, 96, 101–106
    application of, 104–106
        examples. *See* Lopatin's method, examples of
        to paleotemperature, 106
        to reservoirs, 104, 105
        to tectonics, 106
        to thresholds and deadlines, 102, 104
        to timing, 105, 106
    calculation of, 101–102
    correlation, 103
        with R_O, *102*, *103*
        with TAI, *102*, *103*
    interpretation of, 102
Turbidite channels. *See* Migration, factors affecting, turbidite channels

U

Uinta Basin, 22, 27, *35*
Ultrasonic extraction of bitumen. *See* Bitumen, extraction of
Ultraviolet spectra of porphyrins. *See* Porphyrins, analysis of
Unconformity, effect on geologic reconstruction, *98*, 99, 133, 134
Unpaired electrons. *See* Free radicals

Uplift, effect on geologic reconstruction. *98*, 99
Upwelling, *17*, 18
Urea adduction of *n*-alkanes, 60–61
UV. *See* Porphyrins, analysis of

V

van Krevelen diagram, 22. *See also* Tissot diagram
Visual kerogen analyses. *See* Kerogen, visual analysis of
Vitrinite, 22, *23*, 45
   oil-generative capacity of, 69, *70*
   reworking of, 46, 73
Vitrinite reflectance, 45, 46
   application in source-rock evaluations, *67*, 72–74, 86
     examples, 77–82, 107, 108, 113–115
   correlation
     with TAI, *72*, *102*
     with TTI, *102*, *103*
   problem with, 72, 73
Volumetric calculations, 68, 75–77, 79, 80, 131

W

Walnut hulls, as contaminants, 59, 80
Water, loss of, from kerogen, 36, 43
Water of compaction. *See* Migration, factors affecting, compaction water
Water washing. *See* Petroleum transformation, water washing
Wax, plant, 15, 21, 22, 25. *See also* Petroleum composition, waxiness
Wetness. *See* Gas, wetness of
Williston Basin, 92, 93
Wood, 10. *See also* Kerogen type, woody; Trees

X

Xylene, 7

Z

Zeolite molecular sieves, 60
Zooplankton, *12*. *See also* Plankton, Phytoplankton